Mr. Daniel E. Green
2017 Stampede Dr
Spearfish, SD 57783

AF709436

ANTHRACITE RAILROADS & MINING

In Color

VOLUME 1

by

CHUCK YUNGKURTH

To access our full library *In Color* visit us at
www.morningsunbooks.com

MORNING SUN BOOKS INC.
ROBERT J. YANOSEY, President

Published by
Morning Sun Books, Inc.
9 Pheasant Lane
Scotch Plains, NJ 07076
Printed in Korea

Library of Congress
Catalog Card No. 2009932499

First Printing
ISBN 1-58248-288-8

ACKNOWLEDGEMENTS

Books of this type would not be possible without acknowledging the efforts and cooperation of the photographers. A special note must be made of the contributions of Mike Bednar, an author and retired Lehigh Valley employee. He got into places where virtually no one else ventured and recorded them for all of us. Other major contributors were Dr. Art Peterson and Jim Kostibos, who faithfully recorded the anthracite scene while it was still viable.

One of the real joys in examining the slides that Bob Yanosey sent for this project were the views at Wanamie by Dr. Art Peterson and others. While I was quite familiar with the Wanamie operation and have quite a few black and white negatives of it in my collection, these were the first color views I had ever seen. They made a magnificent detailed center piece on anthracite mine railroading for this volume. A note of appreciation is made to F. Charles Petrillo, who came through with a fine view of the Wanamie breaker.

And then there are those fabled old-timers like Robert F. Collins, Dave Augsberger, the fabulous steam shots of Frank Watson, Bill Ellis, as well as the Bennett Collection at Steamtown.

And my wife, Mary, was a constant help in proof reading and spurring me on to complete this work almost on schedule!

Chuck Yungkurth

ABOVE • *A Reading 2-10-2 is working hard leading a mixed train north of Tamaqua on one of the difficult rail routes built to move anthracite from the mines. The difficulty of the grade is shown by the smoke plume of the pusher engine required on the relatively short train.* (Don Ball Collection)

ANTHRACITE
RAILROADS & MINING
In Color
VOLUME 1

ANTHRACITE RAILROADS & MINING *In Color*

VOLUME 1

The most common coal found all over the world and in almost every state of the Union is bituminous or "soft" coal. Anthracite or "hard" coal is found virtually nowhere in the United States except for the Northeastern corner of Pennsylvania. It is scarce all over the world, but some with somewhat similar high carbon content has been mined in Wales. There are many types, ranging from lignite to anthracite with no clear cut divisions between classifications, depending on the amount of carbon, hydrogen, sulfur, ash, etc. Some anthracite was mined in Rhode Island and a significant amount at Crested Butte in Colorado. It has a hard, shiny black surface with a glassy appearance and was sold mostly for home heating applications. The properties of burning with virtually no smoke, leaving little ash and dust when stored in bins made it ideal for home heating.

In the early years after the Revolutionary War wood was the fuel used for home heating. The primitive iron-making furnaces used charcoal made from timber cutting, moving farther back into the mountains as forests were cut down for fuel. Bituminous coal was used by blacksmiths and others but had to be imported from England or brought by sea from some mines in Virginia. Almost limitless amounts of coal in Western Pennsylvania were locked behind the then-impenetrable wall of the Appalachian Mountains. Exposed veins, or outcroppings of anthracite or "stone coal," were known to exist in the mountains on the upper regions of the Lehigh, Schuylkill, and Susquehanna River valleys, but it was regarded as worthless since it didn't burn. Then, it was discovered that anthracite had to be heated to near red heat before it would ignite.

The initial driving force for the development of anthracite mining was the cost and shortage of wood for fuel. Another significant event was the development of iron-making furnaces that could use anthracite to replace charcoal. This led to the establishment of the huge iron- and steel-making industry in the Allentown-Bethlehem and Scranton regions. Earliest transport of anthracite was done by shipping in "arks" down the Susquehanna, the Schuylkill, and the Lehigh, which led to the construction of canals. The arks were a one-way vessel broken up and sold for timber after making the downstream trip loaded with coal. The demand and weather limited use of waterways and the rapid development of railroad technology made canals obsolete soon after they were completed. By 1860 the railroads that followed the same river valleys as the canal had essentially taken over anthracite transport.

Were it not for anthracite mining, most of Northeastern Pennsylvania would be very thinly populated today; much like the present-day Northern Central part of the state west of the anthracite fields.

ABOVE • *Tall breakers silhouetted against the sky in remote locations were the symbols of the anthracite industry. This is the Coaldale Breaker near Tamaqua.* (Jim Kostibos)

Anthracite Fields of Northeastern Pennsylvania and Principal Rail Routes

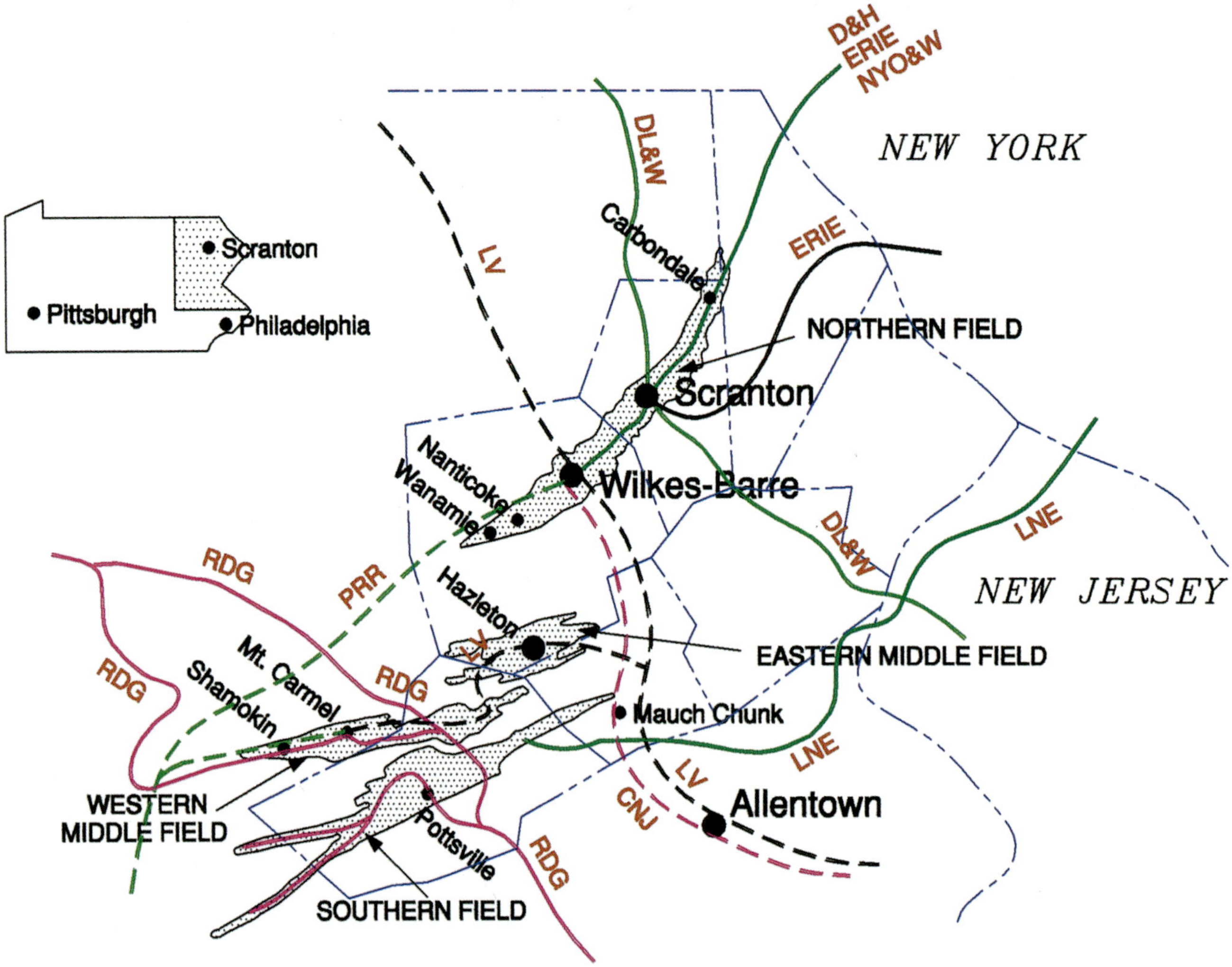

The region is composed of rugged, mountainous land with little top soil and just a small amount of farmable land in the few larger valleys, providing little inducement for settlers. The westward population movement of settlers from New England in the early 1800s bypassed the area, following more attractive routes to the flat, arable lands of Ohio and the Midwest.

The anthracite deposits in Pennsylvania were divided into four major areas referred to as "fields." The Southern Field extended from the Lehigh River at present-day Jim Thorpe, west to near the Susquehanna River above Harrisburg and included Tamaqua and Pottsville. The Middle Field was broken into two parts, with Hazleton being the major city in the eastern half. The western part went from Mahanoy City to Treverton, again almost at the Susquehanna. The Northern Field covered the wide valleys of the Susquehanna and Lackawanna rivers and extended from just below Nanticoke to Forest City above Carbondale, encompassing the major cities of Wilkes-Barre and Scranton.

By mid-nineteenth century it was common for a single business enterprise to be involved in mining, transporting, and retail marketing of anthracite. Since the mines could not exist without the rail transport and there was little reason to build railroads in the sparsely settled regions other than for moving coal, this business organization was entirely logical. But by the late years of the nineteenth century a few huge companies were in complete control of the industry and related railroads, all controlled by New York City financiers. They formed a true monopoly, covering the means of hauling the coal to market, setting prices, limiting the supply of empty coal hoppers to smaller mine operators, owning all property and stores in mining towns, etc and with close links to state and local governments. The most notable of these were the Reading Coal and Iron Company and the Lehigh Coal & Navigation Company. The railroads or their predecessors that were built by these mining interests specifically for the anthracite trade were:

- Reading
- Lackawanna
- Delaware & Hudson
- Lehigh Valley

Other railroads not originally built by the mining interests but which entered the anthracite region by absorbing smaller lines, building extensions or branches into the area, or were taken over by mining interests were:

- Central Railroad of New Jersey
- Lehigh & New England
- Erie
- New York, Ontario & Western
- New York, Susquehanna & Western
- Pennsylvania

These huge rail/mining companies were tremendously profitable and at the beginning of the twentieth century their stocks and bonds were extremely profitable investments and paid huge dividends.

By 1900 the Federal government had cited these coal and railroad companies with restraint of trade and monopolization of the anthracite industry. Among the charges made were excessive freight rates imposed on small, independent coal operators, not supplying sufficient cars to load for movement to markets, establishing wages, and setting market prices. Bitter strikes and prolonged work stoppages against these companies cut the supplies of anthracite for long periods and required Federal intervention. Justice Department court victories led to the companies having to divest themselves of the mining operations and form separate corporations during the 1920s, although for the most part the same railroads served the same mining operations. Some of the new mining companies and their associated railroad were:

Company / Original Owner

- Glen Alden Coal Co. / Delaware, Lackawanna & Western
- Hudson Coal Co. / Delaware & Hudson
- Reading Anthracite / Reading
- Susquehanna Coal Co. / Pennsylvania
- Pennsylvania Coal Co. / Erie
- Lehigh Valley Coal Co. / Lehigh Valley
- Old Company's Lehigh / Lehigh & New England

Nearly all the home heating sales were by small, independent coal yards that had a few delivery trucks. Always a seasonal business, many of the collieries closed down for the summer months and all the railroads stored empty hopper cars on every available yard track and siding. Most of the coal dealers changed their trucks over to making home deliveries of ice for the many iceboxes that lasted well into the 1940s.

The number of railroad and mining company employees was tremendous. In addition to the obviously large number of train operating crews, there were huge car and locomotive repair and maintenance shops at places like Reading, Scranton, Packerton, Sayre, and Ashley. Locomotives were also built at some of these facilities, notably by the Reading. Similarly large office staffs were necessary to handle car accounting, purchasing, and warehousing of supplies. Coal companies operated in much the same manner with company shops, power generating plants, and offices for marketing.

The decline of the anthracite industry was well underway by the onset of the Great Depression, having reached peak output in 1919. Demand went up dramatically during the Second World War but fell drastically soon thereafter. Within the next ten years the anthracite industry was essentially gone. The reasons are many and complex but centered on overproduction resulting in falling prices combined with competition from bituminous coal and other fuels, especially in home heating after WW II. The convenience and attractive cost of oil and natural gas made the switch from anthracite very attractive. Ironically (or maybe typically) the local residents, miners, and coal operators chose to ignore the falling market and while they converted to oil in their homes they steadfastly believed that the anthracite business would come back while mine workings went from five days a week to four, then to three, and finally closing for weeks at a time.

The anthracite railroads fared somewhat better than the coal companies in that their decline took many more years. Some of them, like the Lackawanna and Lehigh Valley, were able to continue some trunk line general freight traffic, such as between Buffalo and the Atlantic seaboard points. But the decline of anthracite mining and the general industrial climate of the Northeastern United States, combined with the impact of the Interstate Highway System and the St. Lawrence Waterway Project, led to the demise of these railroads and ultimate consolidations, such as the Erie-Lackawanna. Then they became part of the massive reorganization caused by the formation of Conrail.

ABOVE · *A typical retail anthracite dealer's storage bunker is pictured at Cobleskill, NY on the Delaware & Hudson. From here the fuel was delivered to the homeowner's coal bins.* (Morning Sun Books Collection)

ABOVE · *Camelback locomotives designed to burn waste anthracite were unique to the Anthracite roads as demonstrated by this Ontario and Western 2-8-0 at Scranton in July of 1947.* (Robert Fillman)

At present, there is some resurgence of anthracite production through strip mining of veins close to the surface. More is obtained through reclamation of the small sizes of coal that were discarded along with the rock and slate into the huge mountains of "culm" that dominated the landscape. Technical developments in coal-burning power generating stations have made it possible to burn waste containing even low percentages of anthracite by pulverizing the material to such a fine level that it is injected into a boiler and burned in the same manner as gas or oil. A few of the large collieries have been reopened to supply this market so the future may be better for the Anthracite Fields.

RECOLLECTIONS

I was a young child in Scranton during the Great Depression. Although we lived on a shaded, pleasant residential street about a half-mile from present-day Steamtown, anthracite coal and railroading was close by and seemed to permeate every aspect of everyday life. One of my earliest school memories was when I was in the second or third grade and was trying to spell and pronounce "anthracite." No matter how I tried it, it seemed to me it should be pronounced and spelled "anthracratic." My teacher was trying to explain to us the annual huge Anthracite Parade that was to be held that weekend on Lackawanna Avenue.

Although crossing Cedar Avenue and going just one city block allowed me to see the huge Glen Alden's Baker Colliery with its monster culm pile just across the Lackawanna River, I can't remember knowing anyone who actually was a miner or worked for a coal company.

The people I knew were craftsmen such as printers, bakers, machinists, cabinetmakers, and painters as well as office workers and administrators.

But mining was ever-present in the background of every day life. The morning and evening news on the local radio stations always included the mine schedule of the operations working the next day. Mine operation was always very sporadic and depended on the season and market conditions, with rail transport of the anthracite similarly affected. During the summer months, every railroad yard and available siding were crammed with empty hopper cars, awaiting the fall upturn in coal demand.

(Dr. Art Peterson)

Most of the mining was done in outlying smaller towns up and down the valley, but there were several large collieries within the city limits such as the Baker, Marvine, the Diamond Shaft, and the Pine Brook. All of these were very close to the center of the city.

At home we could occasionally hear the miners working under the house at night. Not blasting, but the chinking of picks against coal and rock. It probably was workers repairing track and equipment rather than actual coal removal, since that had been done many years before.

Into the north side of the city there was a huge coal stripping operation that looked like the pictures I had seen of the Grand Canyon. South of the city were large flat areas covered with silt marked by dead trees, looking like pictures I had seen of France during World War I.

During World War II, I had a perfect view from my junior high school of the Lackawanna's Bloomsburg Branch trains and of an 0-8-0 switching mine run coal hoppers into the Baker. The double track D&H main that also carried Erie trains was even closer and provided much of the rail action. Also visible was the CNJ roundhouse and track along the Lackawanna River. Glimpses could be had of trains going up the Lackawanna's elevated tracks leading to the passenger depot. And finally, from the eastside windows of the school, slow moving trains on the Erie & Wyoming Branch of the Erie could occasionally be seen crawling along the side of East Mountain. For me, it was not an environment conducive to academic learning and I was constantly in trouble for "window gazing," as it was termed in the notes teachers sent home with me.

My rail expeditions covered virtually every foot of track in the Scranton area and extended to Pittston and Wilkes-Barre by riding the electric Laurel Line. This permitted hiking to the Lehigh Valley's Coxton Yard, Wilkes-Barre depot, and the Pennsylvania's roundhouse in South Wilkes-Barre. Trips to visit family friends in Reading gave me the opportunity to spend hours watching the Reading railroad at Belt Line Junction in steam days. Driving through Tamaqua filled with steam locomotives and smoke was the high point of our auto trips during the late 1930s.

The only line in this book to which I had no personal exposure to during steam years was the Lehigh & New England. Going off to Penn State gave me the opportunity to witness the last operations of Pennsylvania steam at Altoona. By the time I left Penn State in 1953 and moved to the Binghamton area, all the steam was gone except for what little remained on the PRR Elmira Branch.

FOR ADDITIONAL READING

There are numerous Morning Sun Books titles devoted to the individual Anthracite Railroads.

But, the best book devoted to the entire anthracite industry as a whole is **The Kingdom of Coal** by Miller and Sharpless, published by the University of Pennsylvania Press in 1985. While currently out of print, copies can be found with a little research.

Another fine book, with a more corporate viewpoint this time, is **The Story of Anthracite**. It gives a good history of the industry, mining, and breaker operations. This was published in the 1940s and is also out of print, but is in many libraries and may be available to purchase over the Internet.

ABOVE · *By 1956 the decline in anthracite sales was obvious and this sign in Coaldale exhorted the miners to increase productivity.* (Jim Kostibos)

ABOVE · *Reclamation of usable anthracite from the huge mountains of culm, or breaker waste, became common by World War II. Mahanoy City, November 1956.* (Jim Kostibos)

READING

The Schuylkill Canal was opened in 1819 between Pottsville and Philadelphia. It was soon hauling a huge amount of coal, loading at Pottsville at the edge of the Southern Anthracite Field. By 1831 a rail line was opened between Tamaqua and Port Clinton to transport coal to the canal. This was one of the first railroads chartered in Pennsylvania and a portent of the rapid construction of the anthracite lines. In 1842 the Philadelphia & Reading opened its rail line, paralleling the canal for the entire length and the Schuylkill Canal immediately began a disastrous decline due to the severe price competition and year-round operation.

With shops and operations based at Reading, the company soon absorbed most of the smaller lines in the Southern Field, along with huge holdings of mining properties. The mining operations and properties were set up in 1871 as a subsidiary company, The Philadelphia & Reading Coal and Iron Co., which marketed its coal as "Reading Anthracite." Along with the other anthracite railroads, the Reading was forced to divest itself of the mining and marketing holdings by a Supreme Court ruling in the 1920s.

The Reading billed itself as "The World's Largest Anthracite Carrier" and it indeed was. In 1949 it loaded more than twice as many carloads of anthracite as its nearest competitor for the title, the Lehigh Valley. Its alternate routes and many branches covered Southeastern Pennsylvania in a complex network extending out from Reading like spokes of a wheel, reaching Philadelphia, Allentown, Bethlehem, and Harrisburg. The Harrisburg line extended to an important connection with the Western Maryland at Shippensburg. Another spoke was the original P&R route to Pottsville and the entry to the anthracite fields.

An important main line went from Philadelphia to Bound Brook, where it used Jersey Central trackage into the CNJ terminal to accommodate New York City passengers.

At Port Clinton on the line to Pottsville, the original rail route to Tamaqua went on to East Mahanoy Junction, where the railroad wandered through the mining regions over Locust Summit and through Shamokin to Sunbury. Here, it crossed the Susquehanna and then followed the west branch of the river to Williamsport and a few more miles to important connections with the New York Central and the Erie at Newberry Junction. Both of these carriers used the Reading as their routing of cars to Philadelphia. By 1896 the Reading had leased the Catawissa Railroad, which went from East Mahanoy Junction west to West Milton, which was 27 miles east of Williamsport, thereby bypassing the grades and congestion of the route through Shamokin.

BELOW • It is appropriate to start a photographic tour of the Reading and related anthracite here, since this was the site of the original mining operations that led to construction of the Philadelphia & Reading. In this March of 1961 view, Baldwin road switcher #531 heads a passenger train for Reading and Philadelphia. While passenger service was maintained as far as Shamokin and to other points in the Southern Anthracite Field, Pottsville was essentially the termination point for much of the Reading's passenger service north of Reading, as indicated by the size of the train shed. *(Arch Kantner)*

ABOVE · By 1972 the train shed at Pottsville was gone and much of the passenger service was handled by Budd RDCs (Rail Diesel Car), such as the one at the left. Photographer Al Holtz identified the RDC as Train #1060 running from Pottsville to Reading. The caboose is painted in the then-new "banana and avocado," as some railfans dubbed the yellow and green scheme, that was applied to the company's second generation diesels. Road switcher #489 is still in the road's ultra-conservative dark green original diesel livery. *(Al Holtz)*

READING
489
92928
READING

ABOVE · While a fair amount of coal still came through Pottsville, the major traffic moved through Tamaqua and went on to the original P&R route at Port Clinton Junction, fifteen miles south of Pottsville. The Reading Anthracite Company's St. Nicholas Breaker near Shenandoah was one of the newest collieries served by the Reading. Built in 1963 it operated until 2003 when operations were transferred to the company's Girard Breaker near Pottsville. St. Nicholas was typical of latter-day anthracite operations where mine run coal from various shafts and slopes was moved on common carriers, such as the Reading, to large central collieries. At the colliery, the cars were unloaded, the coal processed in the breaker, and reloaded into cars for shipment to the customer. *(Jim Kostibos)*

ABOVE · Reading offset side 55-ton twin hopper #64890 is representative of the typical anthracite coal car of the hard coal roads from the mid-1930s. Only the Lehigh Valley, the Pennsylvania, and the Ontario & Western did not use this American Association of Railroads recommended standard design or some variation of it. The larger 70-ton hopper cars were not preferred by most of the anthracite retailers and industrial users since they did not want to purchase the coal in large quantities. Often their coal trestles and facilities could not handle the heavier cars. The date of the photo is January 1968. *(Rail Data Services Collection)*

RIGHT · The Pacific type 4-6-2 was about the standard power for main line passenger service on the anthracite roads, aside from the Lackawanna, which relied on Hudson and 4-8-4s. The train of standard heavyweight cars is headed by one of the Reading's home-built G-3 class 4-6-2s. Built in 1948, the twelve engines in this group were the newest Reading steam locomotives and were constructed in the company shops at a time when most other lines had already gone to diesel power on passenger trains. *(Frank Watson)*

BELOW · The Reading purchased EMD FP-7 1,500-horsepower unit to replace the Pacifics for main line passenger service. These were conventional F-7 A-units that were lengthened by the builder to accommodate train heating steam generators and the necessary water tanks. Although conservative in black with jade green bands, the scheme used on all Reading F-units was quite handsome. The eight FP-7 diesels were delivered from 1950 though 1952. *(Rail Data Services Collection)*

ABOVE · FP-7 #902 with sister unit heads up Train #7, THE SCHUYLKILL, at Shoemakersville on April 4, 1953. THE SCHUYLKILL was a Philadelphia to Shamokin train that left at noon and arrived in Shamokin at 3:47 PM. The Reading had several trains a day each way between Philadelphia and Shamokin with reclining chair cars and a diner as far as Pottsville on the morning and afternoon KING COAL trains. This 139-mile run was the Reading's longest passenger trip during the 1950s, going on from Pottsville to Tamaqua and then to Shamokin via Shenandoah and Mt. Carmel. *(Bob Wanner)*

ABOVE · A back-to-back pair of FT units head up a mixed freight passing through Schuylkill Haven, which was just a short distance south of Pottsville. An important anthracite branch started here and served many mines, extending all the way to Lykens via Tremont, with spurs to collieries near Good Spring, Glen Carbon, Minersville, and others. The Reading's FT units were the company's first road diesels and were delivered in 1948 just after they built thirty 4-8-4s in the company's Reading shops. *(Jim Kostibos)*

ABOVE · The Reading was one of the most faithful customers of the Baldwin Locomotive Works, carrying on a relationship that went back to the roots of both institutions. Here, AS-16 road switcher #539 works at Schuylkill. It was one of a group of thirty-five 1,600-hp engines that worked all over the system. Unfortunately, the Baldwin products proved to be unreliable and difficult to maintain, resulting in being the earliest first generation power to be replaced. *(John Dziobko, Jr.)*

ABOVE · Class K1-SB #3019 is at Tamaqua on October 9, 1949, ready to head south with a train of anthracite. The Reading had ten of these 2-10-2s built in 1931 by Baldwin. Santa Fe-type locomotives were built in large numbers and were used by many railroads in "drag" freight service where speed was not of importance. When operated over about 25 mph, the huge, unbalanced reciprocating mass of side rods, the main rod, and cylinders rapidly destroyed track alignment. Most roads scrapped the Santa Fe-type locos when motive power requirements shifted to high speed freight service.

(William Echternacht)

ABOVE · Port Clinton was named for a point on the Schuylkill canal. The first rail link to the canal was constructed from here in 1831 along the Little Schuylkill River to transport anthracite from the mines in Tamaqua area to the canal boats. In typical Reading fashion, Tamaqua could also be reached by going through Pottsville, but most through freight to Williamsport left the old P&R line at Port Clinton Junction. This spot along the Schuylkill was a favorite for photographers. An A-B set of FT units heads this southbound mixed freight in September of 1955. *(Arch Kantner)*

ABOVE · Taken from the same vantage point, this view shows a two-RDC set at Port Clinton. The Reading had a fleet of Budd-built RDC cars and used them on the many branches. The service from Philadelphia was virtually a commuter service, terminating at Pottsville and some only going as far as Reading. The date is October 19, 1963 and fall foliage covers the mountains along the river. *(Robert F. Collins, Morning Sun Books Collection)*

ABOVE · A final scene in winter at Port Clinton has a pair of EMD GP-7 locomotives followed by three switchers on a train of anthracite. The locomotives are #623, #626, #2773, #2761, and #2779. By 1976, when this was photographed, the railroad had changed to the yellow and green scheme, which included repainting some of the first generation road switchers. *(Richard Taylor)*

260
260
Reading Lines
260

A train of coal and mixed freight powered by four F-7 diesels passes a good number of spectators at Tamaqua in 1962. Judging from the numbers, it would seem that the occasion was one of the Reading's very popular steam powered "Rail Ramble" excursions. Tamaqua was one of the railroad's major yards where much of the anthracite collected at collieries on the various branches was classified and made up into solid trains for the company piers at Philadelphia and other destinations. *(John Dziobko, Jr.)*

ABOVE · On the same day in Tamaqua, a set of Baldwin road switchers is parked in a siding with an interesting array of autos in the background. The number of bystanders indicates they are waiting for the passage of a steam-powered train. More speculation would lead to the conclusion that the Baldwins are to be helper power on the steep grade westward from Tamaqua to East Mahoney Junction. *(John Dziobko, Jr.)*

BELOW · Tamaqua was surrounded by anthracite mining operations and collieries. This abandoned head frame stands over a shaft excavated to reach the anthracite veins below the surface. Loaded and empty mine cars were raised and lowered by cables that passed over the large sheaves at the top of the frame and then went to winding equipment in the building. Workers entered and left the mine by riding the same lifts as the mine cars. Isolated shafts were often situated a good distance from the colliery, with mine cars hauled by small steam locomotives. These distant shafts simplified the underground movement of mine cars and also complied with mining laws that required mines to have more than a single entrance. *(Jim Kostibos)*

ABOVE · Deep in the heart of the Southern Anthracite Field in October 1973, a train of mine run anthracite is shown passing the abandoned brick depot at Minersville. The train is on the spur from Westwood Junction, which was on the branch from Port Clinton to Lykens via Tremont. Two of the Reading's fleet of Alco RS-3s #483 and #484 are the head end power. Description of the Reading's complex trackage in the anthracite fields is difficult since not only were there many branches and spurs, but there were connecting trackage links between these lines. *(Mike Bednar)*

ABOVE · The same train is shown here at Westwood Junction. Two cars of merchandise for local industries are cut in ahead of the coal loads. The Reading had a fleet of eighty Alco RS-3 road switchers that worked well for the many branch and mine operations. A good customer of Alco, the road had FA and FB units, as well as C424, C430, and C630 locomotives, plus a variety of yard switchers. *(Mike Bednar)*

ABOVE · Pusher units #490 and #447 are at the rear of the Minersville coal train passing through Westwood Junction. Pennsylvania rail safety laws prohibit more than 3,000 pusher locomotive horsepower placed behind the caboose, hence caboose #94003 trails. Note the variety of road names on the hopper cars – Western Maryland cars are visible in this photo. A common belief among rail modelers is that coal hauling railroads used only their own hopper cars. While this certainly was true to a large degree on bituminous haulers such as the Norfolk and Western, it certainly was not true on the anthracite roads after World War II when companies were short of cars and had to use any empty cars on the line. *(Mike Bednar)*

RIGHT · Haucks was a station on the uphill grade between Tamaqua and East Mahoney Junction. A westbound train of empty hoppers heads back to the mines in this February 1963 scene. Power is a lash-up of three EMD F-3 road units. The Reading had six F-3 A-units and six F-3 B-units. Their roster of Alco and EMD hood units was small in comparison to the over-150 first generation road switchers made up of Alco RS-3s, EMD GP-7s, and Baldwin AS-16 locomotives, since "cab" units were ill-suited for anything but over-the-road assignments, where switching or reversal of direction was not required. *(Dave Augsburger)*

ABOVE · At CNJ "R" tower in Allentown, one of the Reading's twelve Alco C-630s, followed by a GP-35, heads up a train of anthracite. These C-630 locomotives were among the last Alcos built in the United States before all operations were moved to Montreal. Some of the mechanical and electrical problems that plagued the C-628 series were remedied in the C-630 design, but by then the damage had been done to the company's reputation, with the result that EMD and GE completely dominated the market. The date is May 15, 1971. *(Mike Bednar)*

READING
475

"Plain Jane" RS-3 #475 switches the end of a coal train in the yard at Allentown on May 14, 1971. The two cabooses are in the newer green and yellow scheme. Most of the Alco RS-3 units remained in the original green, with few ever being redone in the newer, more colorful livery. Caboose #92892 has been upgraded by sealing off some of the side windows, with a single window replacing the original two in the side of the cupola. Soon, cabooses would no longer be required, except for some work trains and for switching service. *(Mike Bednar)*

WESTERN MARYLAND
READING
503
92882
READING

ABOVE · At Allentown Yard, cars of coal from Hazleton are pushed over the track scale for weighing in October 1972. Large collieries had track scales that were used by the railroad to ascertain acceptable weights. Samples from each car were checked by quality inspectors for uniformity of size along with freedom from slate and rock before the car was shipped. Cars received from smaller operations had to be weighed at a railroad facility. Note the hump yard at the right.

(Mike Bednar)

ABOVE · Baldwin switcher #65 spots an LNE hopper at Birdsboro. A large steel foundry at Birdsboro produced castings such as truck side frames for the railroad industry and was an important customer for the Reading, as well as a supplier for the company shops. Baldwin VO #660 is a real vintage diesel, one of ten delivered from 1940 through 1942. The Baldwin yard switchers were probably that builder's most successful product. They were rugged and powerful and gave them years of service.

(Calvin T. Banse, Morning Sun Books Collection)

ABOVE · Articulated 2-8-8-0 1823 is seen at Reading on May 20, of 1950. Only the Erie, the Reading and D&H used such locomotives. As pictured, the machine is "simply" articulated, meaning that all four cylinders work at boiler pressure. It was probably built as a Mallet compound, where the center set of cylinders used boiler pressure steam that exhausted to a set of low pressure cylinders at the front. Compounding gave greater fuel economy but less tractive effort than when all cylinders worked at boiler pressure. Most Mallet types were eventually "simpled." *(William Echternacht)*

ABOVE · One of the Reading 4-8-4s, #2102 smokes it up at Excelsior, PA as it heads up one of the very popular "Rail Ramble" excursions run by the company in 1964. Excelsior is east of Shamokin on the line over Locust Summit. The #2102 along with twenty-nine others were rebuilt from class I-10su Consolidation engines in the company shops at Reading. The term "rebuilt" may have been stretching things a bit, since new frames, drivers, lead and trailing trucks and many other parts were used. However, the new 4-8-4s were fine locomotives and gave excellent service at a time when most other railroads were converting entirely to diesel power.

(Robert F. Collins, Morning Sun Books Collection)

ABOVE · On another Rail Ramble, 4-8-4s #2100 and #2102 pass interested spectators and slow highway traffic at East Lewisburg, PA. The date is October 10, 1963. The train is on the track built by the Catawissa Railroad and was taken over by the Reading in the 19th century. The line started at West Milton, crossed the West Branch of the Susquehanna, and headed east through Dansville and on to East Mahoney Junction. There was little in the way of industry and no mining along the route that followed the mountainside.

(Robert F. Collins, Morning Sun Books Collection)

ABOVE · The destruction of the terrain and environment by the anthracite mining industry was truly horrendous, particularly for the first half of the 20th century. Mountains of "culm," the discarded rock, slate and unsaleable small sizes of coal, dominated the countryside. A double-headed Rail Ramble special is seen passing the site of a former culm pile that has been reclaimed, but has already left behind a barren scene so typical of anthracite country. This is deep in the Southern Anthracite Field somewhere between Mahoney City and Shamokin. The date is October 12, 1963. *(James P. Shuman)*

ABOVE · Gleaming in fresh paint, FP-7 units head up a Reading fan trip at Lofty, PA in May 1969. Lofty was on the Reading's line from East Mahanoy Junction to West Milton. This route allowed through traffic to avoid the congestion and heavy grades on the parallel trackage through the mining areas of the Southern Anthracite Field. Expedited interchange general merchandise from the New York Central at Newberry Junction near Williamsport headed for Philadelphia and points south went this way. *(Bob Wanner)*

BELOW · An A-B-A group of Reading F-units is working to haul a general merchandise freight up the stiff grade at Locust Gap. The Reading black scheme used on cab units with the single green band and yellow pin striping reflected the basic conservatism of the company. The painting of yard and road switchers was more drab, what with the all-olive green. *(Robert F. Collins, Morning Sun Collection)*

ABOVE · The Reading was a faithful customer to the Baldwin Locomotive Works, a carryover from years of purchasing steam power. AS-16 road switchers #551 and #554 are part of a large group of these locomotives purchased from 1951 through 1953. Baldwin yard and road switchers were noted for their starting effort as well as their pulling power in low-speed drag service. Unfortunately, Baldwin products required considerable maintenance and shop work to keep them working. Hence, they were the earliest group of first generation diesels to be scrapped or traded in by most railroads. *(John Dziobko, Jr.)*

ABOVE · The backbone of the Reading's Post-War coal hopper fleet was made up of offset side twin hoppers such as the #81600, pictured here on September 26, 1948. Although the Reading constructed many similar hoppers in the company shops, this one was built by Bethlehem Steel Company, a traditional source for Reading purchases. The car carries the red logo, proudly asserting "America's Largest Anthracite Carrier." *(Lawson Hill, Boston Chapter NRHS Collection)*

RIGHT · Here is a Reading 70-ton triple hopper 6000-series car. The Reading alone among the hard coal carriers retained higher capacity cars, although the Pennsylvania always favored them, being primarily a bituminous carrier. This car is pictured at Dewitt Yard in Syracuse, NY in April 1971. By this time the Reading had gone to the slanted billboard-style lettering. *(Rail Data Services Collection)*

ABOVE · The 100-ton triple hopper was more or less the standard hopper for coal hauling starting in the 1970s, although the Reading was the only carrier to have any significant number in anthracite service. The Pennsylvania and the Erie-Lackawanna also had them, but their anthracite traffic was virtually nil by that time. RDG 484268 was an "off the shelf" standard car built by Bethlehem Steel and was purchased by many railroads. The location is Binghamton, NY in June 1984, with minimalist lettering prevailing. *(Rail Data Services Collection)*

BELOW · One of the lesser-known but more important present-day uses of anthracite is in filtration applications for water treatment, chemical processes, and even domestic uses such as drinking water, swimming pool, and fish tank filters. The finely powdered anthracite is marketed under a variety of trade names such as "Anthrafilt." The Reading had a group of covered hoppers of the type pictured that were devoted exclusively to the haulage of powdered anthracite.

(Rail Data Services Collection)

774
774

Much of the Jersey Central aura is represented by "Camelback" 4-6-0 #774 on an excursion in April of 1955 at Winslow Junction, NJ. It was typical of a fleet of locomotives constructed to burn anthracite. Originally designed for mainline passenger service, they were downgraded into the extensive commuter traffic and branch line service out of Jersey City. The #774 was restored and preserved and is now at the Baltimore and Ohio Railroad Museum. The CNJ maintained and used Camelbacks of several wheel arrangements until well after World War II. *(Robert F. Collins, Morning Sun Collection)*

The Central Railroad of New Jersey, commonly called the "Jersey Central," was a major anthracite carrier even though its early roots are in its namesake state. The original company was formed by combining two smaller lines and was not an adjunct of mining interests, unlike the other "anthracite railroads." In the early years, other carriers including the Lackawanna and the Lehigh Valley, routed their anthracite shipments via CNJ since it was the only carrier at the time to connect Delaware River Valley points with the New York City markets. In 1871 the CNJ leased the Lehigh & Susquehanna Railroad from the Lehigh Coal and Navigation Company thereby directly serving the anthracite region. The LC&N was basically a coal mining and canal company that built a rail line along the Lehigh River from Easton to Wilkes-Barre, paralleling their canal as far as White Haven. The L&S used cable haulage on inclined planes to go over the mountains to reach the Wyoming Valley at Ashley in 1843.

By 1866 a thirteen-mile route had been constructed to permit steam-powered passenger and faster freight trains to bypass the planes. Empty coal hoppers were also sent via the "back-track" bypass. The Jersey Central continued to use the original L&S planes to haul loaded cars from Ashley to the top of Penobscot Mountain until 1948, when all traffic was routed over the back-track and the planes were abandoned. Ashley at the foot of the planes was the main yard and locomotive engine in the Wyoming Valley.

In terms of an anthracite carrier, the Jersey Central can be considered a main line extended from Jersey City to Minooka Junction at Taylor just a few miles south of Scranton. The major part of northbound coal and other traffic was interchanged with the Lackawanna at Taylor with trackage rights on the Delaware and Hudson being used between Hudson, just north of Wilkes-Barre to Minooka Junction. There the CNJ returned on its own rails for a few miles to the minimal engine terminal and yard in downtown Scranton. The main traffic to Scranton proper was the limited passenger service and interchange traffic with the NYO&W.

Other important interchanges were with the Reading and the Lehigh & New England near the extensive yards in the Allentown area. Branches serving various collieries extended from the main line in the Lehigh River valley to Pottsville, Hauto, and Upper Lehigh, along with the important Nanticoke Branch that extended south from the important yards at Ashley.

Following a final bankruptcy in 1967, all the Central Railroad of New Jersey trackage and facilities, including the large Allentown Yard, were taken over for operation by the Lehigh Valley. This arrangement continued until the remaining CNJ trackage and the entire LV became part of Conrail.

RIGHT · The Jersey Central's heaviest freight power was the 86 Baldwin-built Mikados delivered between 1918 and 1925. The wide firebox, which necessitated part of the cab being cut away to straddle the boiler, reveals it was intended to burn anthracite. They worked out of Scranton, Ashley, Bethlehem, and other main line engine terminals. This scene is at Lincoln, NJ where the train has stopped for water. Note the unusual water standpipe at the right that is pivoted and long enough to span two tracks.
(William J. McChesney, Morning Sun Books Collection)

Another symbol of the CNJ and anthracite is the massive Huber Colliery adjacent to the railroad company's Ashley Yard. The Huber was completed in 1939 by the Glen Alden Coal Company to replace several older, smaller collieries in the Wilkes-Barre area. It was the one of the largest and last major anthracite preparation facilities to be built. The structures remain standing to the present day and a community effort is underway to preserve it along with all the interior machinery. It is curious that the Glen Alden, derived from and long associated with the Lackawanna, built this huge complex where it was serviced by the CNJ.
(Alfred W. Bennett, Steamtown Foundation Collection)

RIGHT · The Jersey Central's large engine terminal at Bethlehem was also used by the Reading. In this scene the roundhouse, turntable, coaling tower, and power house are visible. A Reading 4-8-4 can be seen just to the right of the coaling tower. The date is September of 1948 and not a diesel can be seen. One clue as to the date of the photograph is seen in the "CRP" lettering on the end of the hopper car of ashes in the foreground. The Central Railroad of Pennsylvania was set up in 1946 in an attempt to escape the burdensome taxation of the CNJ being chartered as a New Jersey corporation. All trackage and facilities in Pennsylvania were transferred to the new corporation, along with almost 3000 freight cars. This legal maneuver failed within a few years and all property reverted to the CNJ.

(Robert Fillman)

ABOVE · Jersey Central RSD-5 #1615 Alco road switcher works the "Heavy Side Hump" of the Allentown yard on March 24, 1972. While the company had ten RSD-4 units, this was the only RSD-5 on the roster; the differences being entirely within the electrical system. Allentown was the major classification yard for the CNJ, and in later years it was also used by the Lehigh Valley after it and the CNJ shared trackage to Wilkes-Barre. It remains an important yard for NS, which routes New York City area trains to it from Harrisburg, then to Reading, on to Allentown and finally to the Hudson River via the former Lehigh Valley main line. *(Mike Bednar)*

ABOVE · Camelback 0-6-0 switcher was part of a group of class B36 built in 1912 through 1915. It is seen here on the Deerfield Branch at Seabrook Farms, NJ in March of 1949. In later years of their service they were used primarily in industrial areas to work the numerous sidings. The engineer is seated on the cab window sill; a good position to gain visibility of hand signals from the conductor and brakemen. Aside from a roof over his head, the fireman on most Camelbacks was exposed to the elements. *(David Cope)*

ABOVE · Fairbanks-Morse H24-66 "Train Master" #2401 is passing near the village of Mountaintop, PA, which is near the summit of Penobscot Mountain. This is the highest point between the valley of the Lehigh River at White Haven and the Susquehanna at Wilkes-Barre. The CNJ depot here was actually called Penobscot and was the site of a yard and enginehouse. When announced by FM, the 2400 horsepower locomotives were quickly adopted by three of the anthracite roads. Between the Reading, Jersey Central, and Lackawanna, they purchased almost half of all built.
(Jack DeRosset, Dr. Art Peterson Collection)

BELOW · Just a few miles from Wilkes-Barre, Ashley was the major yard and engine facility at the northern end of the system. It was located at the foot of the Ashley planes ascending Penobscot Mountain. In this view in January of 1965 an eastbound 97-car freight headed by a set of grimy F units is passing the inbound mine run coal yard of the Glen Alden's Huber Colliery. The train is just starting up the "back track" grade built to bypass the planes. Ashley was also an interchange point with the Erie, which had trackage rights on the CNJ from Plains Junction to Ashley. *(Dr. Art Peterson)*

RIGHT · The planes were part of the predecessor Lehigh and Susquehanna railroad, opened in 1837 to tap the Northern Anthracite Field by transporting anthracite to the canal at White Haven. Loaded cars of anthracite were hauled by cable up the three separate planes by a steam-powered winding engine located at the head of each plane. Cars rolled down a short downgrade from the head of one plane to the foot of the next. At the time they were built, cable-operated planes were an established technology, while steam locomotives were just becoming practical. This view is looking down from the top of one of the planes. The hauling cable lies between the rails, passing over pulleys at intervals. *(Frank Watson)*

ABOVE · Here, the heavily weighted "barney" car at the end of the cable is up against loaded hoppers of anthracite in preparation for pushing the cut of cars up one of the planes. Originally all traffic, including descending empties and passenger cars, went on the planes, but in 1866 the back track was completed to allow only loaded cars to move on them. The back track route up Penobscot Mountain was used for passenger trains and empties until the road went to diesel power in 1948. Use of the planes eliminated the need for helper and pusher engines with multiple crews up the steep back track grade. Closing the planes also removed the need for an enginehouse and yard at Penobscot. Note the uniformity and the different size of anthracite in the two hopper cars. *(Frank Watson)*

ABOVE · The barneys had wheels that could slide along the axles with flanges on the outer side edges. This arrangement allowed the wheels to change gauge and move on to a set of rails inside the normal running rails. Using this mechanism, the barney descended into a pit. Cars being placed to go up the plane passed over the barney in the pit. When the cable was tensioned, the barney came up from the pit and moved against the cars. At the top of the plane the cars rolled over to a slight downgrade and the heavy barney car dragged the cable back to the bottom of the grade as it was unwound at the hoist house. The barney gauge changing pit sounds complex, but it really was simple and worked very well. *(Frank Watson)*

ABOVE · Camelback 0-8-0 switcher #272 is seen working in downtown Wilkes-Barre in June of 1947. Aside from a very elaborate passenger station, the CNJ facilities in Wilkes-Barre were minimal, consisting of a freight house, some yard trackage, industrial sidings and interchange with other railroads. Note the trainmen riding on the foot boards and sitting on the coupler. Footboards were very handy for crewmen riding on the engine during switching operations, and trainmen constantly hopped on and off them while the locomotive was moving. The Interstate Commerce Commission eventually outlawed their use on all locomotives well after diesels supplanted steam locomotives. *(Robert Fillman)*

ABOVE · SW-9 switcher 1092 spots a refrigerator car at Wilkes-Barre on May 27, 1967. The engine is working in the same area as the previous photograph of Camelback #272. The locomotive is in the standard CNJ green with the Liberty Head and still carries footboards. The brick building in the background is probably part of the Stegmaier Brewery. *(Rev. A. Kovacs, Dr. Art Peterson Collection)*

BELOW · In the rush to replace steam power with diesels, hardly any railroad bothered to preserve any actual steam locomotives, although the Jersey Central was an exception. The Pennsylvania deserves credit for saving one of almost every major class, now at the Pennsylvania State Railroad Museum. Of the Lackawanna, Delaware & Hudson, and the Erie roster, not a single engine was saved by the railroad. Several of the Reading 4-8-4s are still in existence. Here, three CNJ engines are being dismantled at a scrap yard near the Pennsylvania's Buttonwood Yard in South Wilkes-Barre. *(Frank Watson)*

ABOVE · The Jersey Central, Lehigh Valley, and the Delaware & Hudson all had parallel trackage and yard track in central Wilkes-Barre. While there were some street overpasses, there were also grade crossings for a few streets that crossed many tracks. At this crossing north of the depot the CNJ provided the gates and watchman various rail lines' crossings. The gates at crossings such as this were usually actuated by compressed air from a manually operated pump in the gateman's shack. This elevated structure was a Jersey Central standard design, identical to the one at Elm Street in Scranton and others. *(Alfred W. Bennett, Steamtown Foundation Collection)*

ABOVE · One of the Jersey Central's workhorse offset side hopper cars was photographed in May of 1970 at Binghamton, NY. The 55-ton capacity car is one of a group built in 1944. Even with the severe shortage of steel during World War II due to construction of ships and military equipment, a large number of freight cars were built in order to carry anthracite for home heating and industry. *(Rail Data Services Collection)*

ABOVE · Caboose #91514 was of the "eastern standard caboose" design, originated by the Reading and used by the Reading, Jersey Central, Lehigh & New England, Lehigh Valley, and Western Maryland. Many were built in the Reading's shops for the other lines and, in some cases, kits of parts were supplied by the Reading and they were erected in other railroads' shops. There were variations, like end windows, on some railroads' cars. The Lehigh Valley's cars can always be identified since they have three rungs on the cast platform steps instead of the usual two as seen here. *(Rail Data Services Collection)*

ABOVE · CNJ Jordan spreader #9300 is stored in the weeds at Wilkes-Barre. Jordan spreaders were built by a company in the Midwest and used by almost all railroads for spreading newly dumped ballast and plowing out drainage ditches. The cars could also be used for snow plowing, where the extendible side wings were very useful. The view here is the rear of the spreader, which is not often photographed. *(Dr. Art Peterson)*

ABOVE · More company service equipment is seen in this view of the CNJ wreck train, stationed at Ashley. The coach is a standard steel chair car formerly used in mainline passenger service, but here used to transport employees to the location of wrecks or derailments. The two box cars are for tool and supply storage and are originally from the group of 500 standard design single sheathed cars assigned to the CNJ by the United States Railroad Administration during World War I. *(Alfred W. Bennett, Steamtown Foundation Collection)*

BELOW · A northbound mixed freight is shown passing the D&H "MO" tower at Mineral Springs just near Hudson, PA. From here, the CNJ used nine miles of trackage rights on the D&H to reach Minooka Junction, where there was a small yard where most cars were interchanged with the Lackawanna. At Minooka Junction the CNJ reverted to its own track for the remaining four miles to Scranton. Mikado #922 heading the train was part of an order of twenty locomotives delivered by Baldwin in 1925. *(Frank Watson)*

RIGHT · An Alco RS3 running short hood first with an EMD unit is seen passing the former D&H passenger depot near Hudson Yard. The date is September 1967. The second locomotive is one of the twelve SD-35 locomotives that the CNJ received in 1965. These were the first of the second generation acquired by the company, to be followed by a group of SD-40s leased from the Baltimore and Ohio in 1967. The B&O had a large part of the ownership of the CNJ, so cars and locomotives were frequently leased from the larger road.
(Alfred W. Bennett, Steamtown Foundation Collection)

BELOW · In June of 1966 the Miners Mills depot was still standing, although it was many years since the CNJ made a passenger stop here. The standard paint scheme of cream and brown is in remarkably good shape. Near here the CNJ Canal Branch ran Past #14 Colliery to Planes Junction, where it connected to the Erie. The Large Pine Ridge Colliery was located adjacent to the Miners Mills deport, but it was serviced by the D&H.
(Alfred W. Bennett, Steamtown Foundation Collection)

RIGHT, BOTTOM · This view taken at Minooka Junction on September 19, 1954 exhibits a lash-up of Baldwin "baby face" cab units led by CNJ 74-79. It's a truly rare photograph, since Minooka Junction was inaccessible except by hiking in and was virtually unknown except to a few native railfans like Bill Ellis and railroad employees. Perched on the hill above the box cars in the train is the Lackawanna's Taylor depot at the end of their large Taylor Yard. The actual junction and D&H interlocking tower is off to the right on the opposite side of the Lackawanna River.
(Bill Ellis)

74-79
JERSEY CENTRAL
LINES

ABOVE · The latter-day Jersey Central facilities at Scranton were minimal. Virtually everything can be seen in this 1965 view taken from Cliff Street. The Lackawanna River and the D&H mainline tracks are out of the view at the bottom of the photograph. The Lackawanna Avenue Bridge and the magnificent CNJ brick freight house with the imposing tower can be seen at the right. In contrast, the passenger depot was a simple wood structure, the roof of which can be seen behind the standing locomotives.

The reason for this simple building is that plans for building a large Union passenger depot serving the CNJ, D&H, DL&W, and the NYO&W never came to fruition. So the Jersey Central simply carried on with the existing passenger depot although the brick freight house had been erected in anticipation. *(Dr. Art Peterson)*

LEFT · Another photograph of the CNJ in Scranton shows the proximity of the downtown business district. There is a small yard with about four tracks on the other side of the concrete viaduct. The other end of the yard was the interchange with the Ontario & Western and there was a very steep grade up to the team tracks and a few commercial customer sidings that were at the same street level as Lackawanna Avenue. The caboose in the foreground is standing on one of the tracks formerly used to store passenger equipment. *(Dr. Art Peterson)*

LEFT, CENTER · The roundhouse and coaling facilities for steam power were situated about a mile south of the Scranton depot, just off Hickory Street where the Roaring Brook flowed into the Lackawanna River. Now it is the site of an athletic field. The engine house had about six stalls and the coaling tower was a tall wooden structure. When the road went to diesel, power engine servicing was moved to the site of the passenger station. All that was needed was sanding equipment, which was provided by a tank atop of an old signal gantry relocated to the site. Locomotive fueling was also done there, although the actual arrangement for doing so is not clear at this late date. In August of 1964 EMD F-3 #57 is awaiting dispatch with a caboose to Taylor to pick up its train. *(John Dziobko, Jr.)*

BELOW · In a September of 1969 a telephoto lens shot from Cliff Street shows GP-7 #1531 at the minimal Scranton "diesel terminal." The original purpose of the large brick structure in the background is unknown, but checking old Sanford maps shows it was never a CNJ property. In some photographs, a covered hopper with locomotive sand is spotted on the siding adjacent to the building. *(Rail Data Services Collection)*

ABOVE · In this final view of the CNJ facilities in Scranton, two locomotives are framed by the piers and girders of Bridge 60 on the former Lackawanna. EL Alco switcher #550 is working the east end of the EL Scranton Yard. It is somewhat of a mystery why the CNJ kept loco facilities and originating crews at Scranton in the later years, instead of just abandoning all track north of Minooka Junction, with engine servicing performed at Minooka. Aside from team track loads and the freight house, there were only one or two customers at Scranton along with the O&W interchange, which ended with the liquidation of that railroad in 1957. *(Dr. Art Peterson)*

ABOVE · 0-6-0 switcher #107 poses on the team tracks at Scranton just to the north of the freight house. An appearance of an 0-6-0 at Scranton was a rarity. During most of the years of WWII and through to the end of steam the switching chores there were handled by one of the CNJ's 4-8-0 Camelbacks. Use of a larger engine was no doubt due to the very steep grade up to the level of the team tracks and the necessity for frequent runs to Taylor to pick up or drop off cars at the DL&W interchange. *(Frank Watson)*

Wanamie Mine

The Glen Alden Coal Company's Wanamie #18 Breaker was located near the end of the Lee Mine spur off the Central Railroad of New Jersey's Nanticoke Branch that extended south from Ashley Yard. Mining had been done at the site since the middle of the 19th Century, with the Glen Alden Coal Company taking over the mine and breaker from the Lehigh & Wilkes-Barre Coal Company in the 1920s. About 1955 the Glen Alden dismantled the Wanamie Breaker but continued to mine anthracite from Slopes 18 and 19 at the site. At the time the breaker was closed there were about 1,200 workers employed at the mine. A mine car dumper and loading facilities were established at the former breaker location to load mine run coal loaded directly into standard gauge hopper cars. The cars were then moved by the CNJ to the coal company's huge Huber Colliery at Ashley, where they were unloaded, whereupon the coal was then separated from rock and slate, cleaned, sized, and reloaded into cars for shipment to markets.

Closing the Wanamie breaker was part of the movement throughout the anthracite industry to consolidate the preparation of the coal at large modern facilities instead of the aging smaller operations scattered at mine heads. Advanced preparation processes used complex, expensive equipment such as patented Chance or other cone cleaners that used both centrifugal motion and water flotation to separate the anthracite from rock and slate. Labor costs were drastically reduced and quality of the fuel market was improved in a very competitive market. The trend to move mine run coal from mine operations to central collieries started in the 1920s, with the large anthracite producing companies erecting new large collieries, and continued through the late 1930s. A few were completed in the Southern Field after World War II.

The mine cars and steam locomotives at Wanamie were typical of those used throughout the anthracite fields. Cars were usually of very crude wood construction, and built in the coal company shops. Dimensions were determined by clearances in mine tunnels and the amount of coal miners could load per day by union contracts. For many years underground movement of cars was done by mules stabled in the depths of the mines, a practice that lasted into quite modern time in small workings. Electric locomotives powered from wires suspended from the roofs of gangways (main tunnels) moved the cars to the bottom of shafts or slopes. Small steam saddle tank locomotives were used on the surface to bring mine cars from outlying slopes or shafts to the colliery, although electrics were occasionally used on the surface. Track gauge at Wanamie was 36 inches, but anthracite mine trackage in the industry varied from 24 inches to 42 inches with no attempt at industry standardization since there was no interchange of mine cars.

RAILWAY

What made the latter-day Wanamie operation unique was that it is believed to be the last narrow gauge colliery mine railroad to be operated using steam locomotives in the Anthracite region lasting until 1967. Otherwise it was a pretty typical small anthracite operation with no breaker. At Wanamie the steam locomotives were used to move loaded mine cars from #19 Slope to the main hopper car loading equipment at #18 Slope, a distance of about a mile and a half. They also performed switching service and movement of props and supplies. Actual haulage of coal from the slopes and movement to standard gauge hopper cars was done by wire rope haulages and gravity. Investment in machinery and capital equipment at small mine openings such as the last years of Wanamie was always minimal since the amount of mineable coal at the site was known and would be exhausted in a limited period. Underground mining operations at Wanamie seemed to have ceased about 1967 but loading of culm into hopper cars reclaimed from nearby banks, done via trucks for haulage, continued.

The village of Wanamie was a typical "mine patch" in that it was a collection of dwellings clustered near the breaker. These small settlements were usually fairly isolated in wilderness areas at a distance from any larger cities with stores and commercial establishments. Nanticoke, ten miles away, was the nearest city to Wanamie. For many years company stores were the only place purchases of necessities could be made in the mine villages. Documentation of Wanamie mine rail operations in black and white photographs is fairly extensive but the color presentation here is the only known collection to be preserved on color film.

BELOW · The breaker at Wanamie is shown here in this 1955 black and white photograph shortly before it was razed and the mining operation was converted to simply loading mine run anthracite. The "mine run" coal was then brought from Slope 18 and 19 mine openings and loaded into hopper cars for movement to the Huber colliery, where it was prepared for marketing. In the following color views of final Wanamie narrow gauge rail operations, many of the small buildings that are seen were still in use. *(F. Charles Petrillo Collection)*

ABOVE • By 1967 the breaker was gone yet many of the brick buildings such as shops and engine house are still in use. Much of the narrow gauge trackage is in use too, but with some revisions for dumping and loading the mine run coal. Most of this scene shows the storage yard for mine props. In the distance can be seen the huge culm bank where waste slate and rock from the breaker was dumped and many years later was reclaimed and processed to salvage the small sizes of anthracite contained.

(Alfred W. Bennett – Steamtown Foundation Collection)

ABOVE • One of the main purposes of the narrow gauge railroad at Wanamie was to move loaded mine cars from Slope #19 to the dumper and loader at the original breaker site, as seen in January of 1967. On the very primitive rail equipment there were no brakes except on locomotives. Speeds were very low and if the locomotive could not stop or slow the train, a workman would run back along the cars and insert "sprags" between the spokes of the mine car wheels. These were lengths of wood about three inches in diameter and 18 inches long, sharpened to a point on both ends. Sprags were also used to keep standing mine cars from moving.

(Alfred W. Bennett – Steamtown Foundation Collection)

ABOVE · Engine #8 is arriving at Wanamie with a train of loaded mine cars from Slope 19 in September of 1966. The line of empty PRR hoppers is waiting to move to the loader that is out of the view and a few hundred feet to the right. The crude wood bumper blocks and hook coupling are visible on the front of the engine. There were no provisions made for taking up the slack in the hook and chain link couplings.

(Alfred W. Bennett – Steamtown Foundation Collection)

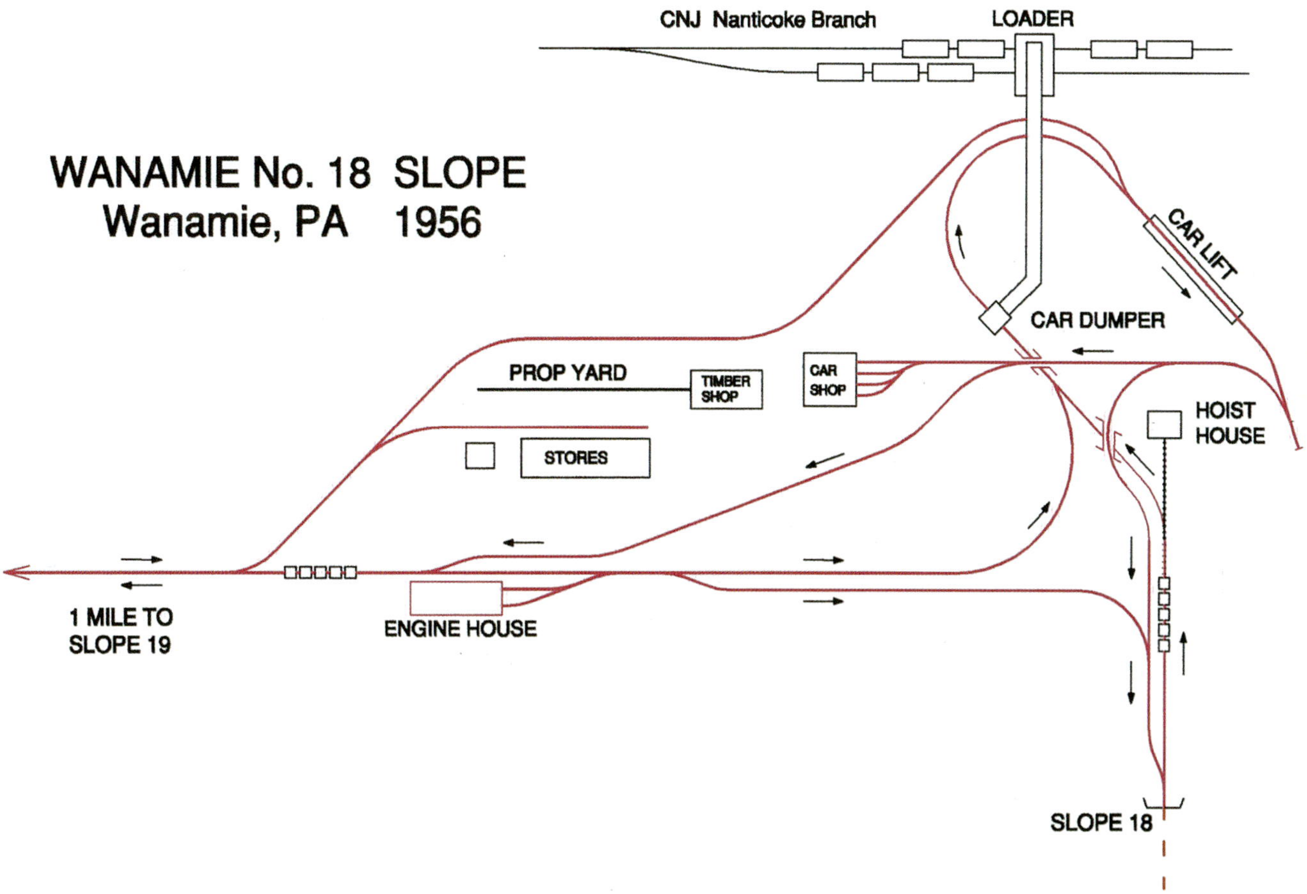

ABOVE · Glen Alden Loading operations at Wanamie; all movements by cable up grades and then down by gravity, except for haulage from #19 Slope and switching in prop yard.

Car movement from #18 Slope was:

- Car pulled up slope to Hoist House
- Cable disconnected, car coasted down slope to spring switch for track to Dumper
- Coasts to dumper
- Car dumped, then allowed to coast down grade to bottom of Car Lift.
- Hoisted up Lift, disconnected from lift chain
- Runs back through spring switch and coasts to return track at Slope

For cars pulled by engine from #19 Slope:

- Cars set off on slide down grade to dumper
- Run through dumper and up Car Lift, as done with cars from #18 Slope
- Switch behind Hoist House set for return track for 19 instead of 18
- Empty cars accumulated on track by Car Shop to return to #19 slope

ABOVE · Loaded cars are getting ready to go down the steep #18 Slope at Wanamie in June of 1965. Cable haulage and gravity were the methods of moving mine cars at the site, except for the haul from #19 and switching in the shops and timber yard. Note the man in the car going down into the mine. Some mines had cars specially built for transporting miners into the underground workings, but at Wanamie and at most other mines the workers simply rode in empty cars. Should the cable or a coupling break, the cars roared to the bottom of the slope, with disastrous results. *(Dr. Art Peterson)*

ABOVE · The electrically-powered large winding drum for the cable of the slope at #18 was contained in this brick building, a remnant of the days of the working breaker. The operator, or "hoisting engineer," was located inside the building and received his signals by bell code or telephone. In the early days, signals were sent by a wire attached to a bell. White markings on the cable let the operator know the position of the cable and the car in the mine. Despite State Mining laws and safety regulations, anthracite mines were extremely dangerous places to work. This picture was taken the same day as the preceding one that shows the car on the slope. *(Dr. Art Peterson)*

ABOVE · From the hoist house, groups of cars rolled downgrade to the dumper. The worker in the view uncouples each of the cars and controls the flow of movement of individual cars into the dump mechanism. The chain and screw eye in the worker's hand seen at the lower left is the fitting used to attach cars to the end of the slope cable. Note that boards have been added to the sides and the ends of the cars. Mine car height and coal loading were determined by the overhead clearance in the workings. Apparently, these cars were originally built for use when roofs were lower than those at Wanamie in June of 1965. *(Dr. Art Peterson)*

LEFT · Continuing with Dr. Peterson's series following the flow of cars in 1965, a car is being emptied within the dump house by tilting. The cars were built with one end having a panel hinged at the top and latches at the bottom for emptying. While an efficient scheme, this meant that all cars had to be oriented correctly with the hinged ends at the same end. The actual dumping mechanism with the dumper shed was quite simple: a short section of track rails pivoted at one end over the pit containing the end of the conveyor to standard gauge hopper cars. *(Dr. Art Peterson)*

ABOVE · The conveyor to the hopper cars on the CNJ Branch was the largest piece of equipment at Wanamie. It was a type very common in handling anthracite, consisting of a sheet steel trough where paddles mounted on a chain drive pushed the coal along. The chain was a continuous one and the paddles on the return section can be seen at the top side of the conveyor. The most common application of such conveyors was to move waste culm to the tops of the banks that were adjacent to every breaker. The date is May of 1965. *(Dr. Art Peterson)*

ABOVE · The end of the conveyor where the anthracite is discharged into waiting standard gauge hopper cars is seen in this interesting shot. In keeping with the hazards at mine workings, the large drive gears are completely unshielded, ready to snag an unwary worker. As mentioned elsewhere in this text, the hopper cars are not owned by the CNJ, in contrast to the popular model railroad view that coal hoppers seldom strayed from their home road. By the 1960s there was a chronic shortage of hopper cars on the anthracite railroads so they maintained their supply by simply retaining any cars that came their way. May 1965. *(Dr. Art Peterson)*

ABOVE · In May of 1965 cars of mine run anthracite are ready for the 12-mile movement from Wanamie to the Huber Colliery at Ashley by the Jersey Central. Cars from the Western Maryland and Baltimore and Ohio are in evidence, which might be expected since there were close corporate ties and traffic interchange between the Reading, CNJ, B&O, and WM. Note the wide variety and highly-piled coal in the cars, which was typical of mine run coal being hauled short distances. *(Dr. Art Peterson)*

ABOVE · The Glen Alden's massive Huber Colliery at Ashley was completed in 1938 and named for the then-president of the company. It was the newest, largest, and most modern anthracite processing complex in the Northern Anthracite Field. The huge glass sheathed structure is still standing as this is written in 2009, complete with internal equipment. It is pictured here in 1979 after being abandoned for some years. A local active group is working to restore the building and internal machinery to an extent that it can be operated as an educational museum and tourist attraction. *(Rail Data Services Collection)*

ABOVE · Back at Wanamie this scene shows an empty mine car rolling from the dumper to the bottom of a cable actuated incline that raised the cars high enough to roll to a short "kick back" reversal track in order to go on to the hoist house of #18 Slope. Here, they will be held until they can be lowered down the slope to mine workings. In the distance can be seen the huge culm dump, later to be removed for processing to reclaim the small-sized anthracite contained. The bleakness of the area is all too typical of worked-out anthracite operations. May 1967. *(Dr. Art Peterson)*

BELOW · The Wanamie loading operation had three Vulcan 0-4-0 tank engines. They were built by the Vulcan Iron Works of Wilkes-Barre, which built all kinds of mining equipment, as well as industrial steam and diesel locomotives. Two engines were used on the trains of loaded mine cars from the #19 Slope, one being used as a pusher. The third was a spare to use while one of the engines was being overhauled or was down for repairs. Here, one of the dust covered engines is switching at the timber yard on January 30, 1967.

(Alfred W. Bennett – Steamtown Foundation Collection)

ABOVE · This undated photograph appears to have been taken after activity had ceased at Wanamie and the engines were not in use. The brick enginehouse is in the background. The engines were typical mine saddle tank "dinkies," with the water for the boiler carried in the U-shaped tank "saddled" on the boiler. Coal was carried in a small bin in the cab. Only one man was used to run the engines, serving as both engineer and fireman. Small industrial engines were built by several companies besides Vulcan, including Porter, Davenport, and the large builders, although Vulcan was the prime supplier for the anthracite region. *(Alfred W. Bennett – Steamtown Foundation Collection)*

BELOW · Wanamie #9 is taking water in June of 1965 at the standpipe near the engine house. The worker filling the saddle tank is probably the "engineer" since one man was essentially responsible for the operation and maintenance of an engine. Note that the steam dome and stack are in openings recessed into the tank in order to stay within the clearances around collieries. Such limitations on the surface were imposed by the trolley wire used where electric mine locomotives were used on the surface. At this date there were no surface trolley operations, but the locos were probably Glen Alden standard designs.

(Dr. Art Peterson)

LEFT · The steam dome, sand dome, and stack located in the opening of the saddle tank are clearly seen in this detailed view. The whistle and safety valve are located on the flat topped steam dome and the sand dome's rounded top is just ahead. Mine locomotives generally did not have bells, which were required by law on standard gauge road locomotives. The water fill cap is barely visible just behind the stack.

(Alfred W. Bennett – Steamtown Foundation Collection)

ABOVE · In June of 1965 the engineer is dumping the ash pan of #5, as can be seen by the falling red hot ashes. Engine servicing facilities at Wanamie consisted solely of the water standpipe. Ashes were simply dumped between the rails and shoveled aside when they piled up too deep. Obviously, coal was everywhere so the engine's tiny coal bunker in the cab was simply filled from the nearest mine car or pile of anthracite. Number 8 in the background is seen with the engineer hanging out of the cab window, giving an idea of the tiny size of the engines. *(Dr. Art Peterson)*

BELOW · Mine cars were generally built by the coal companies to standard designs in their own shops. Most of them were all-wood construction except for hardware, wheels, axles, bearings, and couplings. Even the metal parts were made in company foundries and blacksmith shops. Wood was used because it was not affected by the moisture in the wet coal and damaged sections could be easily replaced. When cars were worn out or damaged beyond repair, the metal parts were salvaged and used to build new cars. These primitive cars were virtually miniature versions of the four-wheeled coal "jimmies" used by the large anthracite railroads until around 1900. *(Dr. Art Peterson)*

ABOVE · Seen here is a typical very low profile electric mine locomotive that was used underground at the Wanamie slopes. The trolley arm is latched down. The operator sat on the tiny seat on the opposite side of the loco near the trolley base, with a small pit for his feet. Needless to say, the operator's head was just a few feet from the live trolley wire that was mounted on the rock ceiling of the mine gangway. The large reel centered on the top of the loco contained cable that allowed the loco to operate a short distance into workings off the main trolley wire equipped gangway. This was done by clamping the end of the cable to the live ceiling trolley wire… another mind boggling safety practice! The equipment on the flat car appears to be a cable winch. The date is September 1970. *(Alfred W. Bennett – Steamtown Foundation Collection)*

ABOVE · Mine props were a vital part of anthracite mining. Major collapses of the entire formation over a mine tunnel or work area were rare. But "falls" were a constant danger. In these incidents a layer of slate a few inches thick at the top surface of the work area falls within a short time after a blasting, much like the plaster ceiling of a room falling in a home. After the blast to loosen the coal the miner checked the "roof" by tapping with a steel bar and listening for hollow sounds that were the signal to the trained ear. If the roof was poor, props made from six or eight inch logs were put in place as supports and were held there by driven wood wedges. This view was taken in June of 1965. *(Dr. Art Peterson)*

ABOVE · The prop yard was a prominent feature of the surface area around underground anthracite mine operations. Tree trunks of proper diameter were delivered to Wanamie and sawed to the proper lengths, as needed. Many collieries bought the props cut to length and delivered them as such, eliminating the need for sawing at the mine. The flat cars in the foreground were used to deliver the props to the work areas. Propping the roof in his work area was the responsibility of the individual miner. It was always a temptation for the miner to skip some of this time-consuming work, but the resulting falls were the chief source of mine injuries and fatalities.

(Alfred W. Bennett – Steamtown Foundation Collection)

LEFT · Here's an electrically-powered crane working on a short section of standard gauge track used to handle logs and finished props at the yard. The crane was built by General Electric. Anthracite mining was a very early convert to the use of electric power, mostly direct current generated at company power plants. The disadvantages of direct current transmission were more than offset by the ease of motor speed control it afforded on breaker machinery, hoist motors, and electric mine locomotives. It also helped reduce the danger of gas explosions underground from open lights. June 1965.

(Dr. Art Peterson)

ABOVE · A carload of long mine props is switched at the end of the prop yard. Two idler flats are used to accommodate the overhang of the long timbers; just like standard gauge railroads used idler flats to handle overhanging loads. A large dump truck of the type used to transport anthracite at strip mining and culm reclamation workings is parked in front of the newly painted Jersey Central hoppers waiting to be loaded for the movement to Ashley. The date is October 28, 1966. *(Alfred W. Bennett – Steamtown Foundation Collection)*

BELOW · With the engineer leaning out of the cab, a train of mine props switches before leaving for #19 Slope on January 30, 1967. The short props are loaded into regular mine cars with one flat of slightly longer props. The mine car ahead of the engine appears in several photographs and seems to be a car of chains and tools moved around with the engine. While there may have been a seat of some kind in the cab, the engineer was almost always seen standing or hanging out of the windows. *(Alfred W. Bennett – Steamtown Foundation Collection)*

ABOVE · Sand was used for a variety of purposes around anthracite mining, the most common method being when it was used on locomotives to prevent drive wheels from slipping when starting heavy trains. Here, a train of four sand cars is being switched near the enginehouse at Wanamie. Note that the sand cars are special low side cars. Railroad cars intended for sand loading generally had very low sides to prevent overloading with the dense, heavy commodity. June 1965. *(Dr. Art Peterson)*

ABOVE · Again in May of 1965 Engine #8 passes the brick stores building as it begins the trip back to Slope #19 with a trainman riding the bump beam. If necessary, he will jump off the moving train and insert sprags between car wheel spokes to help brake it. Chains and tools to be used in re-railing mine cars festoon the front of the engine. With the very poorly maintained and un-sprung cars, derailments were very frequent. *(Dr. Art Peterson)*

ABOVE · The last empty mine car passes the water standpipe while departing Wanamie for the trip back to #19 Slope. The hinging of a car end for dumping can clearly be seen here, including the hinge bar at the top and latching bar at the bottom. Chalk markings were constantly seen on mine cars; these include the word "brake" and what seem to be marks to indicate car count. On the left is a nice collection of vintage autos. *(Dr. Art Peterson)*

LEFT · A Wanamie train with engine #8 between Slope #19 and the unloading facility passes through a rural scene devoid of signs of mining activity and culm banks. Despite the heavy damage and scars created by anthracite mining, nature rapidly covered much of the land with brush and other foliage. Even culm banks would be spotted with clump birch, which seemed to grow anywhere except the very steep slopes. Today, much of the Wanamie is completely overgrown with trees and brush between the standing brick walls of a few structures. *(Alfred W. Bennett – Steamtown Foundation Collection)*

LEFT · Back at Slope #19 in June of 1925 a pusher engine prepares to assist a train of loaded cars to Wanamie. A helper is assisting the engineer by stoking up the fire to raise steam. Normally, the loco engineer handles all the chores on the engine. Chalked on the panel ahead of the cab window is a message exhorting people to get out and vote on a Luzerne County issue. The track can barely be made out in the rock and rubble that has fallen from mine cars. *(Dr. Art Peterson)*

ABOVE · The track between Wanamie and Slope #19 crossed a major road, so a crossing watchman was necessary to flag down traffic when a mine train approached. A typical shanty was built to offer the watchman some protection from the weather, but it had a crossing warning sign that was unique to mine railroads. Pennsylvania laws regulated the shape and size of conventional railroad "crossbuck" and other warning signs and it may have been that there were also regulations for industrial rail crossings. In any event, these curved top signs were common at mine rail throughout the region. *(Dr. Art Peterson)*

ABOVE and BELOW · The Central Railroad of New Jersey's Lee Mine Spur of the Nanticoke Branch passed through Wanamie and extended a few more miles to end at Lee. The importance of the Lee Spur can be seen in the size of the depot at Wanamie. The two-story construction indicates that when originally built, it housed living quarters for the station agent and his family. It has been many years since passenger service existed on the Branch, but before automobiles and good roads became common the CNJ was the only way to reach Wanamie from Nanticoke, Ashley, or the major city of Wilkes-Barre. The building was not in use but in fair condition in May 1967 after being used for the freight agent's office for years after passenger service ended. *(Both – Dr. Art Peterson)*

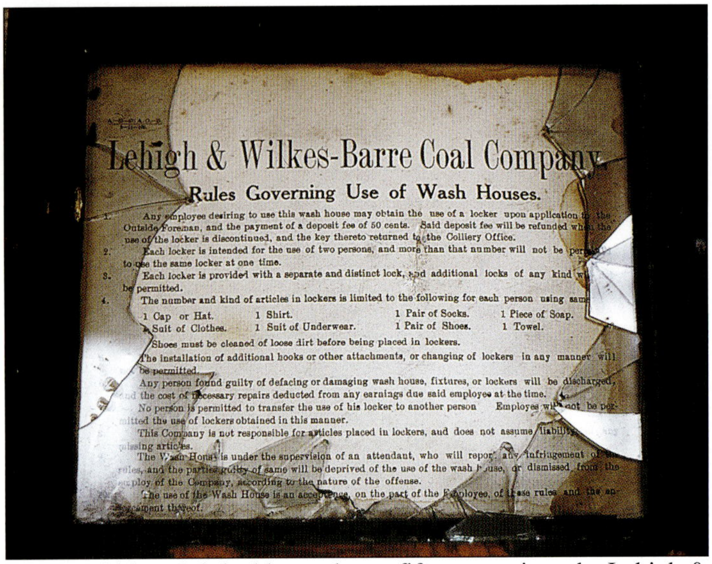

ABOVE · Although it had been almost fifty years since the Lehigh & Wilkes-Barre Coal Company's Wanamie colliery was taken over by the Glen Alden, photographer Alfred Bennett found this sign for employees still posted in the ruins of the Wanamie wash house. The mine wash house was where workers changed into their clothes and gathered their tools and helmets. It was also where they cleaned up after their shift. *(Alfred W. Bennett – Steamtown Foundation Collection)*

BELOW · This unique shot in May of 1967 typifies how, despite modernization and the building of huge facilities like the Huber Colliery, anthracite mining remained very much a brute force activity. Here, a worker moves empty cars at Wanamie. While a locomotive may have been around at the time, it was simpler just to push a car or two around. But likely the steamed up engines were off to #19 Slope. *(Dr. Art Peterson)*

ABOVE · In September of 1972 Lehigh Valley Alco RS units #210 and #211 move empty hopper cars at Wanamie. In that year the LV took over operation of all the Jersey Central trackage in Pennsylvania, including the Nanticoke Branch. Underground mining activities had ceased at Wanamie and operations were limited to reclaiming culm from the waste banks and moving it to working breakers for anthracite. The unique high short hood of the #211 that led to the unit being dubbed "the Hammerhead" can be glimpsed. *(Dave Augsburger)*

RIGHT · Taken in June 1974 this view of a mine car on an incline at Wanamie is labeled "Dumping Coal from Slope at Wanamie." While it is generally understood that underground coal removal had ended some time before this date, it could well be that some intermittent mining was taking place. The mystery remains, as is so often the case in researching anthracite operations and history. *(Jim Kostibos)*

Lehigh & New England

The Lehigh & New England's tangled early corporate history can be traced to the efforts of building a rail connection from Harrisburg to New England via the Hudson River crossing near Poughkeepsie, thereby avoiding the congested and costly routing through New York Harbor. The grandiose scheme of a single railroad company never came to pass but parts of it were built, including the Poughkeepsie Bridge and a railroad going east from Campbell Hall that ultimately became part of the New Haven. Another company that evolved from the tangled complex of proposed railroads built a line extending from Slatington on the Lehigh River to Campbell Hall, NY using trackage rights on the New York Susquehanna & Western and the Erie. This eventually became the Lehigh & New England.

The New Haven's Maybrook Yard just east of Campbell Hall was really the eastern terminus of the L&NE. The railroad lived up to its initial purpose of being part of a route to New England. In addition to that, the company acquired substantial industrial shippers of slate and cement along numerous branches in Pennsylvania, as well as important interchange connections with the Jersey Central and Reading in the Allentown area.

In 1905 the powerful Lehigh Coal and Navigation Company acquired the Lehigh & New England. The LC&N was the organization that opened up the vast anthracite fields west of Mauch Chunk and built the canal along the Lehigh River to transport coal to the burgeoning iron and steel works near Allentown. In 1868 it opened the Lehigh and Susquehanna railroad to move the anthracite and to help with the problems associated with low water and winter freezing of the canal. It also allowed the LC&N to tap into the Northern Anthracite Field. Late in the 19th century the company leased the L&S to the Jersey Central.

ABOVE · At Pen Argyl a fan trip on the DL&W placed Lackawanna 4-8-4 #1641 next to one of the L&NE's Alco FA road diesel units delivered in October 1948. The #1641 was scrapped in 1953, but steam was pretty much gone on the Lackawanna by 1950. So while the photograph is not dated, it was probably taken in 1948. The FA was brand new in its conservative black with white striped livery. *(Gerald Landau Collection)*

RIGHT · The Lehigh Coal & Navigation Company marketed its anthracite under the trade name of "Old Company's Lehigh." The familiar round "fried egg" logo with the red center dates back to a marking used on the company's canal boats and was seen on coal dealers' yards all over the Northeast. The close relationship between the coal company and the L&NE is seen in the same round logo (usually referred to as a "medallion" in the rail industry) used on equipment by the railroad, with only the lettering changed. *(Mike Bednar)*

By 1904 the LC&N acquired the Lehigh & New England in order to reach new collieries in the Panther Valley west of Mauch Chunk. But what really made the L&NE an anthracite carrier was the construction of the "Tamaqua Extension" in 1912, which provided direct access to the LC&N development of coal properties between Tamaqua and Nesquehoning. Using the connection to the NYS&W, the LC&N had a direct route to tidewater for its anthracite.

By 1961 most of the bulk cement-hauling traffic had been lost to trucking and anthracite mining was almost over. This led to the ending of all Lehigh and New England operations in October of that year. Some of the large LC&N collieries at the Hauto end of the line continued to operate. Service to these was provided by the Jersey Central, which took over portions of the former Tamaqua extension trackage as well as other industrial branches.

LEFT · Camelback Consolidation #152 is pictured in steam at the Tadmore engine facility on March 13, 1949. It is representative of much of the L&NE older steam power. The L&NE was one of the roads that used the center cab locomotives right up to the time diesel power replaced steam. The #152 was built by Baldwin in 1911 and was one of two locomotives in the LNE class E-S-13. Tadmor was located between Bath and Nazareth in the heart of the cement producing area. Tadmore yard and engine terminal was opened in 1928 to replace smaller, outdated facilities in Bath. *(The Houser Collection)*

BELOW · The L&NE had quite a large number of heavy 0-8-0 switchers that worked mostly around the cement mills in addition to performing yard switching at Tadmore and Pen Argyl. Pen Argyl was the yard where cars were marshaled into solid trains for forwarding to Maybrook. The #114 was received from Baldwin in June 1915. This photo was taken at Tadmore on March 13, 1949. The #114 was scrapped at Lippsett Steel Products in Coatsville, PA the following year. *(The Houser Collection)*

ABOVE · Danielsville was the point where the Tamaqua Extension started when it was built in 1912. About 12 miles west of Danielsville, the new line crossed the Lehigh River on an impressive deck truss. By 1972 the L&NE had been abandoned and its track removed except for portions taken over by the Jersey Central. The substantial terra cotta block construction is a type used extensively in Southeastern Pennsylvania in the early 1900s but is rarely seen in railroad structures. *(Jim Kostibos)*

ABOVE · The Lehigh Gap was the place where the river cut through Blue Mountain and started across the relatively flat terrain to its junction with the Delaware at Easton. The high deck truss of the Lehigh & New England crossed the double-tracked LV main line, the river, and the CNJ tracks on the west bank. The Lehigh Gap was the only feasible passage for rails through the long mountain that extended for much of the region between the Delaware and the Susquehanna. When the Tamaqua Extension was built, the L&NE had little choice but to construct an expensive bridge to get through the Gap. The impressive span is now dismantled. An LV train of empty hoppers is in the foreground in this 1955 view. *(Arch Kantner)*

ABOVE · Arlington Yard near Tamaqua was a small facility where cars from the various collieries were collected for movement to Pen Argyl. By April 1973 the L&NE had been dissolved as a company and Arlington was part of the trackage operated by the Central Railroad of New Jersey through a subsidiary. Three CNJ locomotives were assigned by the CNJ and worked out of Arlington. Here, a CNJ RS-3 is seen with a cut of hopper cars at the yard in April 1973. *(Jim Kostibos)*

ABOVE · Here, the RS3 is leaving the Greenwood Breaker at Coaldale with a train of loaded anthracite cars on former L&NE trackage. On the mountain above the train is an abandoned breaker. Note that all the cars are from the Reading. The Jersey Central, Lehigh & New England and the Reading had close corporate and operating relationships and frequently supplied cars to each other. Another reason for the mixed ownership of hopper cars on anthracite lines is that the per diem rate on hopper cars was so low that yardmasters simply held empty cars until they were needed, rather than sending them back to the home road. *(Jim Kostibos)*

ABOVE · The Lehigh Coal & Navigation Company is one of the oldest corporations in the United States and is still very much in the anthracite mining and marketing business. It originally started by opening anthracite mining west of Mauch Chunk (now Jim Thorpe) and then constructed the Lehigh Canal in order to move the product to markets. The sign is at the Greenwood Breaker at Coaldale and was photographed in 1971. *(Richard Taylor)*

BELOW · The present-day Greenwood Breaker was erected in the early 1960s by the Greenwood Coal Company on a leased site of the LC&N's former #10 colliery. In 1989 the LC&N took over operation of the new breaker and continues to operate it after a refurbishing that was completed in 2007. It is the only remaining operating colliery in the Panther Valley area and is specially constructed to recover the very smallest-sized powdered anthracite. It is shown here in 1975. *(Mike Bednar)*

ABOVE · This July 11, 1994 scene is the same area shown in the previous photograph. Here, Reading, Black Mountain & Northern engines #800 and #801 work the Lehigh Coal & Navigation Company's Greenwood breaker on former L&NE trackage. The breaker has been upgraded, as revealed by the bright green sheathing. The locomotives are ex-Lehigh Valley that went to Conrail, which were then purchased by the RBM&N and are still in the CR blue paint. The RBM&N went on to become the present-day Reading & Northern, which will be reviewed in Volume II of this series. *(Mike Bednar)*

RIGHT · Lehigh Coal & Navigation Company's GE 44-ton locomotive #125 is pictured at the Greenwood Breaker in 1971. Anthracite collieries generally did not have company-owned standard gauge switchers to move hopper cars. Trackage at the breaker was situated on a slight grade with the empty hopper cars placed on the uphill side by the serving railroad. Then, all car movements by the coal company employees was done by rolling the cars to the loading chutes and then on out to the loaded car tracks. All movements were controlled by the hand brakes on the cars. *(Richard Taylor)*

RIGHT · The same GE switcher is still in use 20 years later at the Greenwood facility but now in solid blue paint and a minimal LC&N logo. The locomotive is said to be "ex-Reading #5133" from the Reading Coal & Iron Company. The General Electric 44-ton locomotives were designed for one-man operation on industrial and short line railroads. Over 300 were built. *(Mike Bednar)*

ABOVE · Here's another view of working loaded hoppers at the Greenwood Breaker in June 1973 on one of the ex-L&NE branches. The Alco RS-3 is apparently one of the locomotives transferred by the CNJ to the subsidiary "Lehigh & New England Railway," which was formed to operate ex-L&NE trackage. The Reading hoppers definitely have seen better days. The #64144 seems to have a covering of plastic to prevent fine anthracite from blowing away. The orange block at the left end of the car meant that it was once assigned to a special service. *(Jim Kostibos)*

BELOW· The entrance to the drift mine at LC&N's #9 mine presents a very typical anthracite scene, complete with miners wearing helmets with lamps. "Drift Mine" meant that the mine entrance was a tunnel that went straight into the mountainside on a level to reach the vein of coal. Electric mine locomotives are used to haul the loaded mine car out of the tunnel and move them to the breaker. Note the low overhead trolley wire going into the mine that was certainly a safety hazard, yet remained standard practice. The electrified mine haulage replaced mules for movement, but mules were used in many small operations through the 1930s. *(Richard Taylor)*

LEFT · In 1869 the Lehigh Coal & Navigation built a tunnel through Locust Mountain near Lansford to enable company-owned trackage at Lansford with the line of the Nesquehoning Valley Railroad at Hauto. The Nesquehoning Valley later became part of the CNJ. When the Tamaqua Extension was built, the Lansford Tunnel allowed the L&NE to reach Hauto and Nesquehoning. The Lansford end of the abandoned tunnel is pictured here in February 1971. *(Richard Taylor)*

BELOW · The Coaldale Breaker of the LC&N, with its huge identifying sign complete with the red and white logos, was shut down by the time this view was made in December 1956. Heaped "culm" is in evidence at the left and in the background and is typical of the bleak landscape. The LC&N dominated the area between Mauch Chunk and Tamaqua, and Lansdale, Hauto, Coaldale, and other communities were virtually company-controlled towns. *(Jim Kostibos)*

ABOVE • On the other side of the Coaldale breaker the track from the chutes for loading hopper cars can be seen, along with one of the LC&N's unique all-steel mine cars standing in the foreground. This breaker is unique. It appears that there were two tracks with cable hauled bunker cars on the incline. All anthracite breakers were tall structures where the mine run coal was sent to the top of the building, where it passed through various crushers and screens, before finally moving on to processes that separated the coal from slate and rock while preparing it for marketing by size. *(Jim Kostibos)*

ABOVE • From one of the walkways at the abandoned Coaldale Breaker the town of Coaldale can be seen in the immediate background. Coal mining villages were usually situated right next to the mines with much of the housing owned by the coal companies for most of the years of operation. When the mines closed there was little in the way of employment, so many workers commuted long distances to places like Bethlehem. Modern breakers were constructed of steel with sheet metal coverings to combat the ever-present danger of fire. *(Jim Kostibos)*

ABOVE · At Nesquehoning, an abandoned head frame stands over a mine shaft of the Panther Valley Division workings of the LC&N. Shafts were often situated a considerable distance from the collieries. This was done to provide more than one entrance to the mine in case of an emergency, as required by law. These laws were passed after disasters where a wooden breaker was located directly over the single shaft to a mine. When the wood breaker caught fire, all air was drawn from the mine resulting in the suffocation of possibly hundreds of mine workers. Additional shafts also reduced the haulage congestion distances from working faces by moving the transport to the surface. *(Jim Kostibos)*

ABOVE · These narrow gauge diesel locomotives were used to move mine cars on the surface at that company's Panther Valley operations. The link and pin coupler, as used on the mine cars, is seen almost at ground level near the foot boards. The photograph was taken at Nesquehoning, PA in December 1956. There was no standardized gauge for mine railroads. The distance between the rails ranged from 30 inches to 42 inches, with each company having its own standards. Since cars never moved between companies' various operations, except for transport by rail or truck, there was no need for standardization. *(Jim Kostibos)*

LNE
LNE
574
571

LEHIGH NEW ENGLAND
584

LEFT · Lehigh & New England steel caboose is one of the railroad's versions of the "eastern standard caboose," as originated by the Reading. In fact, the L&NE's steel cabooses were built in the Reading shops. The date is June 5, 1949 and the location is Tamaqua at what was probably a National Railroad Historical Society event. *(William Echternacht)*

ABOVE · In April 1962 this line-up of cabooses and hopper cars was in storage at Pen Argyl. Operations had ceased in October of the previous year and the equipment pictured is awaiting sale or scrapping. The L&NE had a variety of wood cabooses and a portion of one can be seen in back of the terra cotta building. As late as 1949, the L&NE still had some four-wheel cabooses in use for switching and branch line service. New York State had prohibited four-wheel cabooses many years earlier but the Pennsylvania, Reading, and L&NE used them until after WWII. *(Arthur Angstadt)*

CAMELBACKS ANTHRACITE BURNING LOCOMOTIVES

Early American locomotives nearly all used wood for fuel until about 1860. Bituminous coal was used as locomotive fuel in England but supplies in the United States were far from the seaboard and not readily available. As has been mentioned in the introduction of this book, burning anthracite presented a set of unique problems in that the coal was slow burning and had to be brought to a red heat before it would ignite. For the locomotive designer this meant that a thin, wide fire must be maintained, in contrast to the deep bed of coal in soft coal engines where the grate was placed low between the rear driving wheels.

In 1877 J.E. Wootten of the Philadelphia and Reading railroad invented the "wagon top," or Wootten, firebox that made using anthracite as a locomotive fuel feasible. This design raised the grates above the driving wheels and extended them to the maximum width that clearances permitted. The wide firebox solved the fuel problem but led to difficulties as to how to position a cab of sufficient size for the engineer and yet see beyond the wide firebox. Engines of the period had the firebox extending to the rear of the cab and the cab covered the firebox portion of the boiler. The solution to the problem was to move the engineer's cab ahead of the firebox, straddling the boiler, and provide a small shelter at the back of the boiler for the fireman that lead to the building of many of the "two cab" or "Mother Hubbard" locomotives. Throughout the entire anthracite region these were generally called "Camelbacks" although rail historians designate true Camelback engines as those built by Ross Winans of the Baltimore and Ohio, where a large cab extension was perched on top of the boiler, rather than straddling it. In this book we will use the vernacular "Camelback" designation.

The real attraction of the wide firebox locomotives was that it allowed the burning of "culm" or the unmarketable small sizes from the mines that were considered to be waste. The Camelbacks became the standard power design for the Anthracite railroads. They were built in virtually every wheel arrangement in use around 1910 ranging from tiny 0-4-0 switchers on the Reading to 0-8-8-0 Articulateds built by Baldwin for the Erie. The Lehigh Valley even tried Camelback Pacifics that had a short life before being rebuilt to rear cab engines.

There were some safety issues with Camelbacks, although they were built in large numbers. Few accidents were actually attributed to the cab arrangement that isolated the engineer and the fireman. If anything happened to the engineer the fireman had no way of knowing, although all engines had walkways by which the fireman could crawl forward to the engineer's compartment if he thought necessary. Another perceived risk was that if the main rod fractured, it likely would demolish the engineer's cab with fatal results but again recorded incidents of such failures are rare.

Accommodations for the fireman were nonexistent on the earliest Camelbacks, with the man toiling completely exposed to the elements. Then a small roof was added which gradually evolved to a simple shelter that varied in the degree of enclosure on different railroads. On the Lackawanna another portion of a shelter was added to the front of the tender giving more protection from the elements. Many Jersey Central Camelbacks had sides with door openings on the sides of the rear cab.

Eventually the wide firebox problem and cab location solved itself because by about 1914 locomotives and boilers had become so large that even soft coal burning engines required such large grate areas that fireboxes were extended to the maximum width, main frames were extended, and cabs were placed behind the boiler backhead. The last three Camelback locomotives built were three 0-8-0s constructed by Baldwin for the Lehigh and New England in 1927. Many of the Camelbacks soldiered on through World War II, notable on the Reading, Jersey Central, D&H, NYO&W, and the L&NE.

One final note of interest is that by 1930 none of the anthracite railroads was burning pure anthracite in their engines. The waste culm had found a market in power plants where it was pulverized and burned. So the railroads switched to bituminous from Western Pennsylvania because it was cheaper, although some railroads mixed the two types if prices determined the economy of the mix. Even the "Route of Phoebe Snow" had gone to dirty soft coal, and shortly after World War I abandoned the advertising campaign based on the clean burning anthracite!

LEFT · The Reading had a large roster of 0-4-0 and 0-6-0 Camelback switchers, many of which remained in service until the 1950s when replaced by diesel yard engines. The 0-4-0 engines were used for industrial switching in locations such as the Philadelphia waterfront. Six wheel switcher 1317 is a class B8a and was built by Baldwin in 1907. The crew seems to think it was quite a laugh to be photographed! *(Frank Watson)*

ABOVE • Consolidation 1504 is in switching service at the Harrisburg passenger station. The 1504 is a very typical Camelback, being a medium sized 2-8-0 and was built as heavy freight power but was now relegated to switching service. The I-8a-c engine has been rebuilt with piston valves. Note the air tanks on the top of the wide firebox, a common feature of almost all Camelbacks since there really was no other place to locate them. The engineer is perched on the window arm rest, which seemed to be a favorite position for enginemen during switching operations. *(Frank Watson)*

BELOW • Delaware & Hudson Camelback Consolidation 823 is shown working the Carbondale yard in September of 1947. The D&H was another line that ran Camelback locomotives until the entire road went to diesels. The cab doors open to the walkway linking the front and rear cabs demonstrating that access between the cabs was possible while the engine was moving. The side of the center cab shown had no functional use except as a seat for a head end brakeman or a riding official. The D&H maintained its steam power in immaculate condition down to white tires and running boards on all equipment. *(Robert Fillman)*

LEFT • D&H Camelback 2-8-0 898 stands dead at the Carbondale enginehouse in September of 1947. It is thought that this photograph was taken during a two-day fan trip that toured the DL&W Locomotive Shops in Scranton, Erie facilities at Avoca, and the D&H enginehouse at Carbondale. This engine is equipped with outside ash pans. This was done to comply with an ICC regulation that provisions had to be made to empty ash pans without personnel having to go under the machine. The outside ash pans were applied to a small number of locomotives with wide fireboxes on some anthracite roads but they were not a common solution to the regulation. *(Robert Fillman)*

ABOVE · Another Anthracite carrier's version of a Camelback 2-8-0 is seen in the New York, Ontario & Western's number 216 switching at Scranton in July of 1947. The location is about a half mile from the end of the NYO&W's Scranton Branch where the interchange with the CNJ was made. The embankment in the background is the ramp to the Mulberry Street Bridge and the tops of the spans of the Linden Street Bridge over the CNJ yard can be seen in the distance. *(Bill Ellis)*

RIGHT · Reading 4-6-0 653 is out of road service but is in use to supply steam for heating company buildings at Allentown. It is of class L8se and was built in the Reading Shops in 1908. Next to the Consolidation the 4-6-0 was the most numerous wheel arrangement of Camelbacks constructed, reflecting locomotive building trends of the period. The 4-6-0, or "Ten Wheeler" locomotives were built with large diameter drivers for fast passenger service and with smaller wheels for expedited freight traffic. The 653 is an example of a smaller drivered 4-6-0 as used in freight and commuter service by the Reading. *(Frank Watson)*

RIGHT · Heavy eight wheel switchers were used to move heavy cuts of loaded hopper cars of anthracite at classification yards and at large collieries. At large anthracite preparation plants a locomotive and crew would often be assigned for an entire working shift, the cost of which would be billed to the coal company. Cars of mine run anthracite from outlying mine shafts and slopes would be brought to the large central plant and cars of cleaned and graded coal would be gathered at the outbound tracks. CNJ number 282 has a minimal rear cab and the arched cab window typical of that road's Camelbacks. *(Karl Then)*

LEFT · Ontario & Western Camelback Consolidation 216 awaits a run beside the unpainted enginehouse at Mayfield, PA. Mayfield was the O&W's classification yard and engine terminal for the company's anthracite operations in the Lackawanna River valley. The small drivers and high mounted headlight give the engine a squat, businesslike appearance. The class P locomotive was part of a group of twenty built by Cooke in 1900 through 1904. *(Frank Watson)*

ABOVE · The last group of Camelback locomotives built in the United States was the three built in September of 1927 by Baldwin for the Lehigh and New England. Engines 120, 121, and 122 were basically slightly heavier versions of Camelback 0-8-0 switchers that the road received from Baldwin in 1914 through to 1916. Aside from the two-cab arrangement, the engine is very much up-to-date for the period. Curiously, the L&NE also acquired six much larger conventional rear cab 0-8-0 switchers from Baldwin just a month later in October of the same year. Talking with veteran enginemen has not revealed any dislike of the two cab locomotives and some said they actually preferred them in switching assignments. This view is at Tadmore, PA on March 13, 1949. *(The Houser Collection)*

LEFT · An L&NE Camelback is in steam at Tadmore engine terminal when photographed on March 13, 1949. The end of its service life is near. Built by Baldwin in 1915, the stable of Camelback consolidations was typical of the company's roster of heavy freight power until larger 2-8-0 and some 2-10-0 engines arrived by 1927. The 152 clearly has rebuilt cylinders, new outside steam delivery pipes to the valves, and new outside valve gear. *(The Houser Collection)*

LEFT · Delaware & Hudson Camelback 2-8-0 number 915 is a perfect example of how the problem of putting the engineer's cab ahead of the wide firebox went away. This engine has been extensively rebuilt using the frame of one locomotive and the boiler from another, a common practice on the D&H. The boiler is clearly from a Camelback, as revealed by the spacing of the dome and the bell mounted in the place where the front cab was. A short cab has been placed behind the wide firebox. This photograph is at Carbondale, PA. *(Frank Watson)*

ABOVE · The wide firebox on D&H 160 can be seen extending beyond the cab wall. The D&H, like most of the anthracite roads, did extensive rebuilding of steam locomotives. The 160 was built in 1928 from parts of class E-5 2-8-0s that were built by Alco in 1907 and 1908. The E-5s were built in both two-cab and single-cab versions, with the 160 using the boiler from single cab 1027. It was in service at Carbondale in 1947 and was scrapped in 1952. The D&H was noted for keeping its steam engines in meticulous condition, even applying white paint to the tires, rods, and running boards on the lowliest of yard engines. *(Frank Watson)*

ABOVE · Reading 1687 was one of the company's large fleet of heavy Consolidations that for years was the anthracite burning power for slow-moving coal trains. Baldwin built 75 class I-9sb engines in 1918 through 1921. The 1687 was still under steam on November 7, 1948 by which time the Reading was committed to diesel engines. The rusted boiler front shows lack of maintenance and the anticipated removal from service. The photograph is at the Erie Avenue engine facility in Philadelphia. *(Frank Watson)*

ABOVE · The fifty class I-10sa 2-8-0s were among the heaviest of this wheel arrangement ever built. Only the D&H and the Western Maryland had comparable Consolidation locomotives. Built by Baldwin in 1925 at a time when almost all other railroads were ordering 2-8-2s and 4-8-2s that could handle higher speed merchandise trains, these three coal carriers stayed with the dependable, slow speed power. The notched cab sides allowed a larger cab that straddled the firebox and projected beyond the boiler backhead, giving more space for fittings within the cab. This is at the Erie Avenue enginehouse in Philadelphia. *(Frank Watson)*

BELOW · The Jersey Central received its first 2-8-2 Mikados from Brooks in 1918. Number 870 was part of a second order from the same builder that was delivered in 1920. These and follow-on groups of almost identical locomotives were the largest freight power on the CNJ. The wide firebox and cut away cab sides reveal that the intent was to burn anthracite although the 1920s were the period of transitioning to bituminous fuel on the anthracite. Fortunately wide firebox engines could burn bituminous coal, usually with a change of grates, but the converse was not true. *(Frank Watson)*

ABOVE · The CNJ's last order of steam locomotives for passenger service was the group of fifteen heavy Pacifics received from Baldwin in 1930. They were used on the line's premier passenger trains such as the QUEEN OF THE VALLEY and THE HARRISBURG SPECIAL that ran between Jersey City and Harrisburg, jointly operated with the Reading. Again, the wide firebox indicates a locomotive intended to burn anthracite. The photograph is at Wilkes-Barre on October 10, 1948, probably on a fan trip since appearances of CNJ passenger trains during daylight hours in that time period was a rare occurrence.

(Frank Watson)

ABOVE · Reading Pacific 112 was built in a group between years 1916 through 1923 at the company shops as part of the group of class G1as locomotives. Anthracite culm was still readily available as cheap locomotive fuel during those years. It is shown here at Bethlehem where it has just set off two cars from Philadelphia that will be added to the Lehigh Valley's connecting BLACK DIAMOND. These were the Reading's first Pacifics and the prototype for one of the early HO locomotive kits put out in the late 1930s. The Reading tried to be loyal to coal burning power by building a group of 15 Pacifics in 1948 even as the company was receiving FT diesels from Electro Motive.

(Frank Watson)

ABOVE · Prior to Joseph H. Neulle becoming president of the Delaware and Hudson in 1938 the railroad was operated simply as a pipe line for slow moving shipping of anthracite. Neulle saw clearly that with the declining anthracite business the future of the company was in providing expedited merchandise service from connections at Wilkes-Barre and Binghamton to Canada and New England. To accomplish this the company ordered a fleet of state of the art 4-8-4s and 4-6-6-4s from Alco to replace the aged 2-8-0s that made up the road's entire roster of freight power. But the specifications for the new locomotives specified that they use mixed bituminous and anthracite as fuel, presumably to take advantage of anthracite culm as available. The 1507 is pictured on exhibition at Carbondale in 1951, the same year that the D&H received its first over-the-road diesel power.
(Robert F. Collins, Morning Sun Collection)

ABOVE · The Santa Fe 2-10-2 type was a popular locomotive design in the "drag freight" era where moving maximum tonnage at slow speeds was the primary mode of railroad operation; particularly among coal carriers. But the need for faster trains to move merchandise at higher speeds soon led the railroads back to eight driver engines such as Mikados and Berkshires. The massive amount of reciprocating weight of side rods, main rod, and pistons kept speeds of 2-10-2s below 30 mph or serious track damage resulted. This view of Reading 3010 is at Gordon, PA on August 1, 1950.
(Robert F. Collins, Morning Sun Collection)

BELOW · The final chapter of anthracite burning steam was the construction of 30 4-8-4 type engines at the Reading shops in years 1945 through 1947. Reading records indicate these were "rebuilt class I-10sa 2-8-0 locomotives." It appears that the company used the boilers, cab, cylinders, and various appliances from the old locomotives. These were very successful engines and were even borrowed by the Pennsylvania to ease shortages of power in 1956. The Reading preserved a few of the T1 series and used them on "Rail Ramble" excursions. Some of these are still in existence including the 2124 at Steamtown. The 2111 is seen in spotless condition at Gordon, PA on March 10, 1956.
(Arthur Angstadt, Hawk Mtn. Chapter NRHS Collection)

While the mighty Pennsylvania is not usually considered an "anthracite road," it had a significant presence in the region. Car loadings of anthracite originating on the PRR were generally about the same magnitude as that on the Erie, which is traditionally listed as an "anthracite road." So while the origins of the PRR were certainly not based upon the anthracite transport, any discussion of railroads in the Northeastern corner of the state must include the PRR.

The most significant Pennsylvania route in the anthracite region was the major branch extending from Sunbury to Wilkes-Barre. It ran along the east bank of the Susquehanna and paralleled the Lackawanna's Bloomsburg Branch that was on the west bank. At Buttonwood Yard just south of Wilkes-Barre it connected with the D&H at Hudson by means of the Wilkes-Barre Connecting Railroad, jointly owned by the PRR and D&H. In terms of interline traffic the D&H was virtually an extension of the Pennsylvania Railroad to Canada and New England, particularly for bituminous coal from the western part of the state.

While not serving a large number of collieries, the mighty PRR did extend a few tentacles into the anthracite fields. A short branch from Sunbury to the LV at Lehigh Valley at Mount Carmel was a vital route for iron ore shipments from the Great Lakes to Bethlehem. It also serviced a large mining operation near Shamokin and several smaller ones. The branch was the remnant of the projected Sunbury and Pottsville Railroad of the 1850s that was never built beyond Shamokin. The Schuylkill Valley Branch wandered from Philadelphia through Pottstown, Reading, Pottsville, and Hazleton, and eventually connected to the Wilkes-Barre Branch at Nescopeck. Trackage rights on the

Lehigh Valley between New Boston Junction near Delano and Tomhicken were used to make this a through route. Until about 1940 a Wilkes-Barre to Philadelphia passenger train operated over the route.

A short but important branch extended south from Nanticoke to the Susquehanna Coal Company's workings at Glen Lyon. The PRR had strong financial interests in the early years of the coal company, which was part of the Hanna Coal Company and eventually the huge Berwind-White interests. Just north of Harrisburg, the Lykens Valley Branch reached a few anthracite operations in the western tip of the Southern Anthracite Field. An interchange with the Reading was made at the end of the branch.

BELOW • Although an interchange connection was listed in South Wilkes-Barre, the prime route for the Pennsylvania at Buttonwood Yard and the D&H facilities at Hudson was done using the jointly-owned Wilkes-Barre Connection Railroad. This bypass route crossed the Susquehanna near Buttonwood and then went through Plymouth and Kingston. It then went back across the river to the D&H yard. The connecting railroad was also used by the D&H to reach the important Loree Colliery and other Hudson Coal Company workings near Plymouth. Here, Alco RSD-15 #6816 is leading a train arriving at Buttonwood in November 1967. *(Rev. Al Kovacs, Dr. Art Peterson Collection)*

ABOVE · After delivering a coal train to the LV, GP-30 #2216 and three other units are on Lehigh Valley trackage in South Wilkes-Barre while returning to the PRR engine facilities in September 1967. The Pennsylvania's roundhouse and servicing area were located about a mile north of Buttonwood Yard in a congested industrial area, dating from before the yard was built. The company's passenger trains went beyond the enginehouse for another two miles to reach the Lehigh Valley passenger depot, where they terminated.

(Rev. Al Kovacs, Dr. Art Peterson Collection)

BELOW · Buttonwood Yard was taken over by the Penn Central in early 1968. The yard was abandoned by the PC after a major flood during the 1970s. When the PC became part of Conrail, the entire ex-Wilkes-Barre Branch to Sunbury was taken over by the D&H (it is currently part of the Canadian Pacific). GP-9 #7190 is shown here switching the north end of Buttonwood in June of 1966. Steam operations at Wilkes-Barre ended on October 18, 1953 and the PRR Consolidations used for switching were replaced by GP-7s and 9s.

(Rev. Al Kovacs, Dr. Art Peterson Collection)

ABOVE · Alco produced the 2,500 horsepower C-425 between October 1964 and December 1966, so the #2426 must have been virtually brand new when switching in downtown Wilkes-Barre in June 1965. The Pennsylvania was a latecomer to dieselization and once they decided to retire all steam the builders were flooded with orders. Henceforth the company placed diesel orders to whatever builder would take them, resulting in a roster that had some of almost every type of diesel built during the period. *(Rev. Al Kovacs, Dr. Art Peterson Collection)*

BELOW · A few miles south of Buttonwood the PRR had a mine branch that went from Nanticoke to serve the Susquehanna Coal Company's Glen Lyon Colliery. Shown here in 1974, the breaker is a very old wood structure that has been sheathed in corrugated steel. At one time the Susquehanna Coal Company was part of the PRR's holdings and was their foothold in the anthracite industry. While the railroad divested control of the company, they retained a close working relationship throughout its convoluted corporate history. *(Jim Kostibos)*

ABOVE • Although a bit shabby with weed-grown trackage, the Susquehanna Coal Company was still in production in 1968. The loaded Berwind hopper cars indicate that, at the time, the colliery was being operated by the Berwind-White Mining Co., another business with strong Pennsylvania Railroad connections. Berwind is a vast mining company with many operations in Western Pennsylvania and West Virginia. At Hollidaysburg near Altoona, they once maintained a large car repair operation to maintain their fleet of privately-owned coal hoppers. *(Alfred W. Bennett, Steamtown Foundation Collection)*

RIGHT • *(Jim Kostibos)*

ABOVE · This Berwind hopper is a PRR class two-bay GLa type design. These were built in huge quantities by the railroad starting in 1905, and in 1949 the company had over 25,000 still in service. Some of these were built for Berwind-White, while others were sold to the coal company after being in service on the PRR. Close examination of the photograph revealed that #4584 was one of those built in 1905, making it 65 years-old when snapped in Binghamton, NY in 1970. *(Rail Data Services)*

BELOW · The Berwind-White hopper car fleet seems to have been absorbed by the Penn Central and given BCX reporting marks. But here is BWCX #10968 at East Allentown on May 14, 1975 in fresh paint. The car is a Pullman-Standard PS-3 design and was probably a New York Central car. The reemergence of the BWCX reporting marks is unexplained and the car is not listed in any *Official Railway Equipment Registers*. Nor are the BWCX reporting marks listed as being in use in 1975. It could be that the cars are to be used by Berwind strictly on company trackage and are simply being delivered. *(Jack DeRosset, Collection of Dr. Art Peterson)*

LEFT · Pennsylvania Consolidation #1261 was switching at Pottsville, PA in 1952. Pottsville was on the Philadelphia to Hazleton line that passed through Reading. The PRR was definitely a minor player in the Schuylkill Valley, which was dominated by the Reading. Hazelton was pretty much a Lehigh Valley town, but the PRR maintained its trackage through the region until the Penn Central merger. The PRR was never very much inclined to build locomotives for switching; preferring to use downgraded mainline power, such as heavy Consolidations in that service. *(Bruce Kantner)*

ABOVE · The Glen Burn Colliery on the PRR near Shamokin is a site where mining started in 1843. The first breaker was erected in 1857 with successive structures built over the years. This view was taken in July 1980, the last year that the breaker was used to process coal reclaimed from culm banks. For a short time mine tours were conducted, as the site was promoted as a tourist attraction. Opening of the nearby Ashland Mine Museum by the State of Pennsylvania led to the demise of the community-based effort. *(Rail Data Services Collection)*

ABOVE · Mine run anthracite was carried from the drift mine opening to the breaker by means of all-steel mine cars; in contrast to the crude wood cars used by Hudson Coal Co., the Glen Alden, and most other anthracite operations. The Glen Burn used 42-inch gauge mine track and electric locomotives. The long, covered car at the left was used to carry miners to the underground work faces. Since the overhead trolley wire for the locomotives was simply fastened to the mine roof and was very low, the roof was an important safety feature in keeping the seated miners from contacting the wire. *(Jim Kostibos)*

ABOVE · The waste slate, bone, and small sizes of coal that were separated from saleable anthracite within the Glen Burn breaker were hauled to the top of the "culm" dump by means of rail cars on a cable-operated incline. A car can be seen about halfway up the incline. The massive culm pile was nearly two miles long and dominated the north city of the village of Shamokin. One of the problems with the culm banks was that they often caught fire due to spontaneous combustion. These fires were very difficult to extinguish and often led to residents being forced to abandon their houses, under which poisonous gases were being generated by fire. A rust-colored patch of ash marks a burned area. *(Jim Kostibos)*

ABOVE · In August of 1980 the huge culm bank from Glen Burn operation can be seen behind the ex-Reading switcher and Conrail caboose at Shamokin. The Reading trackage closely paralleled the Pennsylvania's from Shamokin to Sunbury, where the route crossed the Susquehanna and went on to rejoin the company's other route to Williamsport. The switch, with points and operating stand in the center of a bridge, is unique and shows the measures sometimes taken to reach industries. *(Rail Data Services Collection)*

RIGHT · Shamokin acknowledged its landmark culm bank with this roadside sign along Pennsylvania Route 61. For decades the culm banks were regarded as a useless eyesore, which they certainly were. But coal-burning boiler technology developed the ability to burn powdered coal as if it were petroleum or natural gas. In fact, some power plants are able to switch between these fuels in a short time, depending on price fluctuations. The Pennsylvania culm piles are disappearing as the usable anthracite in them is recovered. A great deal of it is burned in nearby cogeneration plants, which supply heating and electrical power for buildings. *(Jim Kostibos)*

BELOW · PRR #220986 is a Class H-31c car photographed at Old Bridge, NJ in 1965. The Pennsylvania made 70-ton hoppers their standard, starting as early as 1910. But in 1941 a small group of the H-31 twin hoppers with conventional vertical side stakes were built, presumably to meet the requirements of small coal users who wanted only 55-ton shipments, such as retail anthracite dealers. These replaced some of the aging GLa cars like Berwind #4584, shown on page 99. Five hundred more cars of Class H-31b were built in 1942, of composite wood and steel construction, thereby saving a small amount of scarce wartime steel plate. After the WWII steel shortage passed, about half of the H-31b cars were rebuilt with steel plate replacing the wood planking. These were designated Class H-31c. *(Dr. Art Peterson)*

ABOVE · The last series of hopper cars built by and for the Pennsylvania were the class H-43, of which 7,500 were built starting in 1964. There were subclasses with minor variations within this group. The 100-ton cars were basically designed for bituminous service but were used to load anthracite from Glen Burn in the final years of that operation. The #225128 was at Binghamton, NY in June 1964. The yellow disk painted on it indicated the car was assigned to special unit train service; a marking usually ignored. *(Rail Data Services Collection)*

RIGHT · PRR #269562 is one of the small group (for the Pennsylvania) of H-36 triple hoppers built at the Altoona Shops in 1957. They were a transitional design between the company's long tradition of quad hoppers to the mass-produced triple bay H-39 class. Photographs of these rare cars are seldom seen since they were lost in the company's monster fleet of coal hoppers. *(Rail Data Services Collection, Dr. Art Peterson)*

BELOW · A Consolidation is switching while in mine run service just a short distance north of Shamokin. The date is February 22, 1957, very close to the end of all steam on the Pennsylvania. As seen by the engineer's position, the engine is backing with a cut of hoppers and one of the PRR's classic N5 steel cabooses. These cabooses were built starting in 1914 and created quite a stir in the railroad news since they were the first all-steel cabooses ever constructed. *(Bill Ellis)*

LEFT · Another one of the Pennsylvania's N5 series cabooses has been rebuilt by adding steel I-beam collision posts on the end platforms. First built in 1914, they were the first all-steel cabooses to be constructed and were the railroad's standard design for almost fifty years, being built with slight variations throughout the period. Thanks to the fact that they didn't leak, many lasted well into the Conrail period and were preferred over all the other myriad designs that CR inherited! This photograph is at Cadiz, OH in August 1968. *(Rail Data Services Collection)*

ABOVE · One of the Pennsylvania's huge I-1 Decapod type 2-10-0s stands at Mt. Carmel water tank. The branch from Sunbury to Mt. Carmel was one of the last stands where these engines were used to move solid trains of iron ore east from Sunbury to the interchange with the Lehigh valley for movement to the Bethlehem Steel works at its namesake city. The line was on the steep grade up to Mt. Carmel and multiple locomotives were required to handle the ore trains. *(Bruce Kantner)*

ABOVE · An A-B-B-A lashup of EMD F-units coupled to an I-1 was snapped standing at Mt. Carmel in August 1957. F-3 #9500 was one of the very first EMD road diesels purchased by the PRR and was built in 1948. The 2-10-0 Decapods were built in large numbers by the company and performed faithfully for many years, although they were considered an obsolete design when adopted by the PRR. The Pennsylvania was very reluctant to abandon steam power, especially after spending a huge amount on the development and construction of a series of advanced steam locomotive designs. *(Arch Kantner)*

RIGHT · In 1952 a Baldwin switcher handles a mixed freight in a heavily wooded valley location on the Schuylkill Branch. Without being able to see the locomotive number, exact identification is not possible, but it is probably a Post-War 1,000-horsepower S-8. The locomotive builder was notoriously inconsistent in model types and locations of identifying details. *(Bruce Kantner)*

BELOW · In this view at Lofty, PA a Pennsylvania Railroad excursion train of modernized coaches is seen on Lehigh Valley trackage. The PRR line between Pottsville and Nescopeck used trackage rights on the Lehigh Valley for the 20.9 miles between Tomhicken and Laurel Run Junction. This trackage was used by the PRR to reach Hazleton, where the company used the LV depot in the years of passenger service. Trackage in the Middle Anthracite Field was a tangled web of parallel lines, trackage rights, and mine branches between the RDG, the LV, and the PRR. *(Bruce Kantner)*

216
LV
216

MIDDLE FIELD

The Lehigh Valley was formed in 1855 as a direct competitor to the Lehigh Coal and Navigation Company's canal and mining operation monopoly. It was a completely rail venture intended to consolidate a few lines connecting mining operations west of the Lehigh with Easton. Early coal shipments were transferred to the Central Railroad of New Jersey in order to reach the Hudson. By 1855 it had extended to Mount Carmel, where connections were made with western lines. In 1867 it reached Wilkes-Barre by a circuitous grade on the side of Penobscot Mountain. It soon became one of the major shippers of anthracite from the Wyoming Valley and reached Waverly, NY from where it connected to Buffalo using trackage rights on the Erie. In 1884 the LV completed its own right-of-way to Buffalo.

As an anthracite carrier, the Lehigh Valley dominated the eastern half of the Middle Field with a network of branches and alternate routes. The western half was shared with the Reading and the Pennsylvania. Carloads of coal were yarded at points such as Delano and Hazleton before being forwarded to junctions on the main line in the Lehigh Gorge. Eastbound traffic moved to the major facilities at Packerton, while loads for western destinations went to Coxton Yard near Pittston.

Like the other anthracite roads, the Lehigh Valley was forced to divest the mining operation by court decision, leading to the formation of the Lehigh Valley Coal Company, which operated many mines and collieries in the Eastern Middle and Northern fields. In this volume the complex network of LV mine branches and spurs in the area centered on Hazelton and extending to Mount Carmel will be pictured. A few of these mine workings lasted into the Conrail years and some to the present-day, served by the Reading and Northern operating over former LV and RDG trackage.

LEFT • In 1972 the LV took over providing service to the CNJ's customers in Pennsylvania, since the CNJ's trustee had received permission to cease all rail operations in that state. As part of that service agreement the Lehigh Valley acquired the CNJ's yard and facilities at Allentown, which had been modernized soon after WWII. This allowed moving most operations from Packerton to Allentown and closing the small yard at Easton. Alco #216 assisted by EMD switcher #181 is pulling a cut of loaded hoppers up the hump and across the car retarders at Allentown in October 1973.

(Mike Bednar)

ABOVE · The Lehigh Valley's premier passenger train, the BLACK DIAMOND, pauses at Bethlehem on March 7, 1959. Note the carts of mail and express waiting to be loaded. Loss of the bulk mail business to air and trucking was a major factor in the loss of passenger train service on the anthracite roads. The Alco PA locomotives powered main line passenger trains to the end. As the number of trains was reduced, they showed up in various freight assignments, notably the Sayre-Binghamton trains connecting with the D&H. *(John Dziobko, Jr.)*

ABOVE · Lehighton and Packerton were close enough to be almost considered one yard. Train crews worked out of Packerton and engine servicing was done there. Here at Lehighton the wrecking train is stored in readiness for emergencies. Covered hoppers awaiting overhaul at the Packerton car shop are stored near the muddy Lehigh River in February 1965. Packerton was the major yarding point for all anthracite loads collected on the branches in Middle Anthracite Field. Eastbound coal loads could be sent by a connection from Ashmore near Hazleton to Tannery on the main line near White Haven. *(Dr. Art Peterson)*

ABOVE · The Jersey Central over the LV main line at Lehighton was a favorite spot for rail photographers. A set of Alco FA units passes under the bridge in this March 1959 view. The Alco FAs were generally regarded as poor performers with serious engine problems. The LV received ten FA and ten FB units in 1948 as part of the move to turn all mainline freight service over to diesel power. All were rebuilt during the 1950's, but by 1972 all were off the roster. *(John Dziobko, Jr.)*

BELOW · Baldwin switchers #137 and #235 work the east end of Packerton yard in June 1965. Baldwin's diesel yard engines were the most successful in the builder's line of diesel locomotives. They were rugged and dependable and many lasted almost to the end of the Lehigh Valley. The #137 is a 100 hp war baby model VO and was built in 1944. The other one of the pair, #235, was delivered in 1950 and is a 1,200 hp S-12. *(Dr. Art Peterson)*

ABOVE · The crew is ready to board a westbound train at Lehighton, headed by EMD F-3 #524 with a B unit and two Alco RS-11s assisting. From here to the top of Penobscot Mountain near Wilkes-Barre it is an uphill climb. It's June 1966 and by this time the Pennsylvania had reluctantly taken control of the LV and a number of RS-11s were supplied by the PRR to ease the shortage of power. Such mixtures of various locomotives were common, as the LV struggled to have enough working units to keep the road running. *(Dr. Art Peterson)*

BELOW · Pushers in the form of a pair of Alco C-420s work against three cabooses on the rear of the train leaving Lehighton seen in the previous photograph. At this date the fifteen C-420s were among the newest locomotives on the railroad. Unlike the C-628s, the 420s were considered to be reliable engines and went on to serve on the D&H after the LV became part of Conrail. *(Dr. Art Peterson)*

ABOVE · The spectacular Alco C-628s in their black and white scheme were the all-time favorite LV diesel locomotive among railfans and modelers; with possibly the PAs as the other contender. Unfortunately, these engines were the design that pretty much sealed Alco's fate as a builder. They were plagued with tracking problems traced to the unique tri-mount arrangement for truck pivoting. The #629 and #631 are seen at the Packerton engine servicing facilities in August 1971. *(Dr. Art Peterson)*

ABOVE · Newly rebuilt covered hoppers stand ready for service near the Packerton Car Shop. While building the locomotive and car shops at Sayre, PA in 1904 did much to centralize locomotive and car work, the LV still retained a car building and repair site at Packerton. In later years the Packerton shop carried out construction and repair work on hoppers for anthracite and covered hopper cars for the anthracite and cement traffic that originated nearby. *(Dr. Art Peterson)*

ABOVE · With the virtual collapse of the anthracite industry from the 1950s through to the 1970s, hopper cars on the Anthracite railroads virtually disappeared. In 1970 the Lehigh Valley had less than 600 hoppers in service. One result of this downsizing was that color slides of hopper cars from anthracite carriers are almost impossible to find since photographers did not "waste" color film on them because they were now scarce. The #25710 is assigned to salt service and lettered to return to Ludlowville, NY. Hauling rock salt in open top cars is unusual, but not unheard of, in times of cover hopper shortages. *(Morning Sun Books Collection)*

ABOVE · The valley of the Lehigh, Schuylkill, and Susquehanna rivers provided the first entries into the southern anthracite regions, with canals providing a transportation system. The prime early entrance to the eastern end of the Southern Field was at Mauch Chunk in the steep-walled Lehigh Gorge. The view here is from Flagstaff Mountain, just south of Mauch Chunk, the present-day Jim Thorpe. A northbound Lehigh Valley train is passing a cut of stored empty gons on the siding. The Jersey Central main line can be seen on the opposite bank at the right.

(Arch Kantner)

ABOVE · This is thought to be the eastbound Lehigh Valley's BLACK DIAMOND with an Alco PA, as seen from across the river at Jim Thorpe in 1950. Yard tracks in the foreground belonged to the CNJ. For roughly a century the two competing railroads had duplicate trackage along the Lehigh River to White Haven and then over the mountains to reach Wilkes-Barre in the Wyoming Valley. It was not until both lines were bankrupt and in their last throes of declining traffic did the companies do the long obvious thing of sharing trackage, eliminating the expensive duplication in 1965. Even today, the Lehigh Gorge above Jim Thorpe is wild and largely inaccessible except by rail, on foot, or by boating. *(Arch Kantner)*

ABOVE · Asa Packer is credited with being the founder of the Lehigh Valley Railroad. Packer came to the Lehigh River and started as a boat builder and soon operated his own anthracite hauling canal boat. His business acumen and other business investments eventually made him a leader in the Mauch Chunk community with involvement in the proposed Delaware, Lehigh, Schuylkill and Susquehanna railroad to haul anthracite to canal boats at the Lehigh River. He gained control of the construction of the DLS&S and changed the name to the Lehigh Valley Railroad in 1853. His baronial mansion overlooking the present-day Jim Thorpe has been preserved, complete with period furnishings, and is open to the public. *(Dr. Art Peterson)*

ABOVE • Motor Car #11 was a gas-electric built by Electro Motive Corporation with a body supplied by Osgood-Bradley. It is seen here in 1946 approaching the Hazleton depot with a three-car train of light weight "trailer" cars constructed specially to be used with the gas electrics. The LV had a good-sized fleet of gas electrics purchased for the many branches in New Jersey, Pennsylvania and upstate New York. *(Frank Watson)*

ABOVE • Being a relatively large city, Hazleton worked to assure it had rail passenger service and aggressively resisted any attempts to discontinue the connection to Lehighton. The LV replaced the gas electric passenger train with a two-car set of Budd built Rail Diesel Cars. They operated between Hazleton and Lehighton until all LV passenger service ended in 1961. The two RDCs are standing in front of the ornate brick Hazleton depot on July 7, 1952. The 26-mile trip to Lehighton took just a little under an hour with scheduled stops at Weatherly and Mauch Chunk with four round trips on most days of the week. A businessman could leave Hazelton at 7:25 AM and be in New York City proper at 11:15; good time even by today's standards! *(Arthur Angstadt, Hawk Mtn. Chapter NRHS Collection)*

ABOVE · Jeddo-Highland #7 Breaker was still in operation and served by the Lehigh Valley in August 1972. At this date there was not an active mining operation here, but the breaker was being used to process anthracite reclaimed from some of the huge nearby "culm banks." Three LV switchers headed by the #262 are working the operation located at Drifton, not far from the colorfully named Pink Ash Junction. *(Mike Bednar)*

ABOVE · Still at Jeddo, the crew working the colliery prepares to leave after supplying the breaker with empty hoppers. An empty box car picked up at local industry is cut in ahead of the newly painted caboose. The breaker is in a sad state and would appear to be abandoned, but a worker can be seen standing at an opening. Culm from the banks was hauled to the breaker by heavy, off-road type trucks. *(Mike Bednar)*

ABOVE • An eastbound train of iron ore is seen near Centralia in 1951. The ore was received in solid train loads from the Pennsylvania, which received it from lake boats. The ore moved over the LV to the huge Bethlehem Steel works at its namesake city and was an important revenue source for the railroad. While hauled in standard hopper cars at this time, the ore was so heavy that the cars were only loaded to about one third of their cubic capacity. Centralia achieved national notice when it was ordered permanently evacuated due to an uncontrollable mine fire directly under the village. Despite the order, a few residents elected to stay. *(Arch Kantner)*

ABOVE • An A-B-B-A set of F-3 units is shown leaving Mount Carmel with an eastbound ore train. It was unusual to use gondola cars for ore but car shortages frequently resulted in using whatever rolling stock was available. In steam years 2-10-2 locomotives working out of Delano engine terminal manned the ore trains. In the background is a bank of the distinctive "pink ash," left after a fire consumed the culm bank. Mount Carmel was the farthest western point of the Lehigh Valley's extensive coverage of the Middle Anthracite Field. *(Arch Kantner)*

CONRAIL

On April 1, 1976 in order to salvage the ailing railroad network of the Northeast, Congress put into effect the Regional Rail Reorganization Act, which created Conrail. This combined the ill-fated Penn Central, the Lehigh Valley, the Erie-Lackawanna, the Central Railroad of New Jersey, the Reading, and the Lehigh & Hudson River into a single corporation supported by federal funding. Hence, with the exception of the Delaware & Hudson, which continued independently, all the Anthracite Railroads went into Conrail. Two of the lines in the group identified with the anthracite trade, the Lehigh and New England and the New York, Ontario & Western, had been abandoned for some time prior to the advent of Conrail.

At the beginning of operation, Conrail identified over fifty shippers associated with the anthracite on its lines, ranging from collieries, mines, coal bagging plants, culm reclamation, and simple truck dumps. Virtually all the trackage in the Anthracite Fields was taken over, along with pruning of duplicate routes, non-productive branches, and unneeded facilities. There was also a massive sorting of cars and locomotives, eliminating the excess and obsolete. It took years to renumber and repaint everything into a uniform scheme, so equipment was seen in pre-merger paint for several years, as will be seen in the photographs in this chapter.

In 1997 the Conrail organization had served its purpose and the federal government divided the routes and assets into two portions that were sold to the Norfolk Southern and CSX. Prior to this, nearly all the Conrail trackage in the Middle and Southern Anthracite fields was taken over by the Reading & Northern. The main through traffic routes in the region are now operated by Norfolk Southern, with trackage rights for the Canadian Pacific, which operates the former D&H. These include the former LV between Wilkes-Barre and Allentown, the PRR branch between Wilkes-Barre and Northumberland, and the Lackawanna trackage from Scranton to Binghamton. Several small regional carriers operate remnants of original anthracite carrier trackage in order to serve local industries.

Even though the sea of blue that replaced the LV's Cornell Red, the Reading's "Banana & Avocado" and the EL's gray and maroon was regretted by rail photographers (though few regretted the passing of Penn Central black), there is no doubt that Conrail saved rail service in the Northeast. Yet now, even Conrail blue is a memory, the company having been divided and sold to CSX and Norfolk Southern.

ABOVE · By now, this location at Allentown Yard is a familiar one. The date is December of 1978, a bit more than two years after the formation of Conrail. By then a good amount of the predecessor lines equipment has received the blue paint scheme. CR #8653 is a former Lehigh Valley unit working the drill tasks at the end of a recently arrived cola train. The previous owner of the "eastern standard" caboose next to the locomotive seems to have been the Reading but the other with port hole windows is clearly a Pennsylvania N5c. *(Mike Bednar)*

BELOW · The Ironton Railroad was a small industrial line jointly owned by the Reading and the Lehigh Valley. As such, it was part of the properties absorbed by Conrail. Here, Ironton Baldwin #751 works a cut of anthracite hoppers being weighed at Allentown on a wintry December day in 1976. The Reading influence is seen in the simple dark green paint scheme. The locomotive number fits in with the Reading's numbering series for 30 DS 4-4-1000 switchers delivered in 1948-49, so the Ironton loco probably was part of the Reading purchase. *(Mike Bednar)*

ABOVE · This is a Conrail coal train at Mine Hill, PA on June 27, 1976, only two months after the CR takeover of the Reading still in its original as-delivered paint and not yet renumbered. Purchases of new diesel switchers were few by almost any railroad in the 1970s but the Reading opted for ten new EMD MP1000C locomotives in 1974. Equipped for multiple unit operation, they were well-suited to operations in the anthracite region, much the same as the Lehigh Valley employed dynamic brake equipped SW-8s for mine runs. *(Mike Bednar)*

RIGHT • By 1978 the motive power from predecessor roads had been renumbered, although many units had not been repainted. Here, deep in anthracite country, a pair of ex-Reading switchers handles a coal train at Good Spring, PA. Originally the site of a large colliery, Good Spring was at the end of a spur off the RDG branch that went to Lykens, where it connected to the PRR. The remarkably well-preserved depot retains the Reading's standard cream and brown paint. Except for the semaphore blades, even the train order signal is intact. *(Mike Bednar)*

ABOVE • The Lehigh & New England's Tamaqua Extension was described in the section devoted to that line. Conrail #9627, still in Reading green, switches in the remnants of the L&NE Arlington Yard near Tamaqua in October of 1978. The abandoned yard buildings are seen in the background scene. The L&NE trackage to New Jersey has been deserted just beyond the structures, with access to the region now through former CNJ trackage to Hauto. Aside from the loss of anthracite shipping, a great many of the area railroad problems can be attributed to excessive branch line trackage and parallel competing routes, which often ran for many miles within a short distance. *(Mike Bednar)*

ABOVE • The identical scene at Greenwood Breaker was shown earlier on pages 77-78 with CNJ power switching the hopper cars. By 1978 Conrail was servicing the breaker with 100-ton hopper cars replacing the 55-ton twin cars so typical of anthracite traffic. The larger cars were an indication that the coal was going to industrial users rather than to small local retail dealers for home heating deliveries. By this date a considerable number of cars had been repainted and lettered for Conrail. *(Mike Bednar)*

LEFT • Working on former Lehigh valley trackage at Beaverbrook, PA, Conrail SW-8 #8678 is noted by the photographer as "stalled again," possibly due to the poor trackage, which seems to end in the foliage. The crew seems to be perplexed by a problem. The #8678 is the former Lehigh Valley #262, still in the Cornell Red scheme, and was one of the switchers delivered with dynamic brakes for service on the anthracite branches. The dynamic brake fan can be clearly seen just ahead of the cab. *(Mike Bednar)*

BELOW • With typical Pennsylvania October fall foliage in the background, a pair of ex-Reading MP1000C still in green hauls a coal train through Westwood Junction in 1978. While there is a motley group of hopper cars in the train, the first one is still in Reading paint. Westwood is on the former Reading line that extended from Port Carbon and Lykens via Tremont. The junction just west of Pottsville is the point where a spur to Glen Carbon started. *(Mike Bednar)*

LEFT • Few of the vast number of twin hoppers so widely used for anthracite carriage survived into Conrail. As a result, photographs of CR twin hoppers are very rare. Like the locomotives stored at Altoona, virtually every unused yard and siding that could be spared was filled with cars awaiting disposition. Offset side hopper #404831 is clearly an ex-Reading car, as can be seen from the blocked out lettering. Photograph location is at Binghamton, NY in August of 1980. *(Mike Bednar)*

ABOVE · Four switchers operating in multiple handle a Conrail coal train passing through a winter scene at Shaft Switches in January of 1979. The scarred landscape remaining after dumping of culm followed by strip mining is made to appear even more desolate by the light coating of snow and an overcast sky. "Shaft" was a station on the Reading. All locomotives have been renumbered into the CR roster but only the lead unit has received the blue paint and CR logo, the others still retain the LV scheme. *(David Augsburger)*

ABOVE · Here is the same westbound Conrail train AH-1 (Allentown-Hazleton, perhaps?) on a different day in February of 1979, led by #8657 at Coalport along the Lehigh River. Coalport was just a short distance north of Mauch Chunk and was an important point for loading anthracite into canal boats. Some of the earthworks of the long-abandoned Lehigh Coal & Navigation Company's canal can be seen as a faint line between the tracks and the river. Interestingly, the lead unit is not one of the ex-LV dynamically equipped switchers. It is probably the leading unit because it has been recently repainted and reconditioned by Conrail. *(Mike Bednar)*

ABOVE · In June of 1978 Conrail train AH-1, with empty hopper cars, passes a track crew working at Hetchel in the wilds of the Lehigh River Gorge. Hetchel was once a station on the Central Railroad of New Jersey and is now where the rail line along the Lehigh passes from former CNJ trackage to those that were once the LV. As has been noted, the Gorge was almost inaccessible for most of the distance between Jim Thorpe and White Haven, except by rail, hiking, or river. Here, an off-road jeep driver has managed to find a route. *(Mike Bednar)*

ABOVE · The Jeddo-Highland #7 operation at Harliegh, PA was being serviced by Conrail in September of 1977 by a pair of multiple unit connected switchers. The Lehigh Valley #253 has not yet been renumbered, since at this date CR was less than a year old and still dealing with the process of sorting out and renumbering locomotives of the predecessor lines. At this time #7 was processing culm from nearby banks to salvage the anthracite contained in the waste from early breaker operations. The operation at Jeddo will be reviewed in detail in Volume II of this book series. *(Mike Bednar)*

Sixteen years later the Jeddo breaker was still standing and served by Conrail. Although appearing to be abandoned, the facility is still in use, although with periods of shutdown. Construction is typical of breakers built during the 1930s and later years using structural steel and covered with corrugated sheet steel. Many breakers of all wood construction lasted through World War II. Fire was a recurring calamity in these structures, often resulting in spectacular blazes. *(Rail Data Services Collection)*

ABOVE · Two Conrail former Reading switchers hastily renumbered, with "CR" stenciled on the blacked-out Reading diamond, work the yard at St. Nicholas, PA in October of 1978. One of the last large breakers built was the St. Nicholas, located near Shenandoah. It was served by the Reading and is pictured elsewhere in these volumes. This was just off the main line that went from East Mahoney Junction through Mount Carmel, Shamokin, and Sunbury, and then rejoined the other part of the Reading at West Milton. *(Mike Bednar)*

ABOVE · CR #9624 is again seen working in the Southern Anthracite Field at Minersville in October of 1978. Minersville was on the branch to Glen Carbon and was one of the larger villages in the region west of Pottsville. Note the Penn Central cars behind the locomotives, reminders of the ill-fated merger of the Pennsylvania and New York Central railroads, which ultimately led to the formation of Conrail to salvage the Northeastern rail situation. *(Mike Bednar)*

BELOW · In September of 1977 Lehigh Valley Alco C628 #639 had become Conrail #6735 and is seen in service at Binghamton, NY. It certainly has had better days, having lost its original "snowbird" paint scheme. Its being in use reflects some of the shortage of good, reliable motive power that Conrail experienced in its first years. Most of the large Alco units absorbed into the CR roster eventually wound up in ore service at Mingo Junction, Ohio. *(Rail Data Services Collection)*

ABOVE • Another large Alco that survived briefly on the Conrail roster was C630 #6769, one of fifteen of this model delivered to the Pennsylvania in 1966 and 1967. This locomotive became Penn Central #6319 after the PRR / NYC merger. This photograph is in June of 1979 and was taken at Sayre, PA. The footbridge over the east end of the yard was well used by rail photographers. *(Rail Data Services Collection)*

ABOVE • Another veteran of anthracite road origin was CR #5975, seen here in June 1976 at the former Pennsylvania's Juniata Shops in Altoona. Sadly, it appears that the ex-Erie-Lackawanna locomotive is being cannibalized for parts. At this time the yard north of the Altoona shops was filled with locomotives of virtually every builder and type from all the railroads folded into Conrail, all awaiting their fate. The #5975 appears to be an ex-Erie locomotive that could have operated out of Avoca, where the Erie's trains of anthracite originated. When the branches to Avoca from Susquehanna and Lackawaxen were converted to diesel power, the new units working the anthracite traffic were mostly EMD GP-7s. *(Rail Data Services Collection)*

ABOVE · Conrail #9245, still in Lehigh Valley Cornell Red, is seen at the former Lackawanna diesel shop at Scranton in June of 1979. When Conrail assumed the vast fleet of cars and locomotives from the predecessor lines, it also acquired several major and many smaller car and locomotive repair facilities. Former Pennsylvania locomotive shops at Altoona for EMD and GE road power continued to be operated by Conrail. For some years Scranton shop handled work on some of the smaller groups of locomotives, such as yard switchers and Alco products. The shop is now a major part of the Steamtown facilities of the National Parks Service. *(Rail Data Services Collection)*

ABOVE · A local train of anthracite ending with a "northeastern standard" caboose passes the abandoned Reading interlocking tower at Pottsville Junction on June 24, 1983. The caboose seems to be of Reading origin but has received some modifications, such as the chest high bar and other changes in the arrangement of railings on the end platform, all of which reveal the Pennsylvania Railroad influence that lasted through the Penn Central and on into Conrail. It can be determined that it was not an LV caboose, since they all had three treads in the cast steps, rather than the two seen here. *(Mike Bednar)*